THE EXPLAINER

From Déjà Vu to Why the Sky Is Blue, and Other Conundrums

THE CONVERSATION

CSIRO

PUBLISHING

National Library of Australia Cataloguing-in-Publication entry

The explainer: from déjà vu to why the sky is blue, and other conundrums.

9781486300501 (paperback)
9781486300518 (epdf)
9781486300525 (epub)

Questions and answers.
Curiosities and wonders.

032.02

Published by
CSIRO PUBLISHING
150 Oxford Street (PO Box 1139)
Collingwood VIC 3066
Australia

Telephone: +61 3 9662 7666
Local call: 1300 788 000 (Australia only)
Fax: +61 3 9662 7555
Email: publishing.sales@csiro.au
Web site: www.publish.csiro.au

Front cover: image by iStockphoto
Set in 10/12 Adobe Minion Pro and Optima
Cover design by Andrew Weatherill
Text design by James Kelly
Typeset by Desktop Concepts Pty Ltd, Melbourne
Printed in China by 1010 Printing International Ltd; reprinted in Australia by Ligare

The articles in this book were originally published by The Conversation.

FOREWORD 1

The idea behind this book is a simple one. Those in the know – academics and researchers with deep subject expertise – share their knowledge in plain language with those wanting to know.

The result: a series of 'explainers' on concepts from the everyday to the curious.

Comprehensive and clearly explained, each explainer gives you credible information and might just spark the occasional 'a-ha' moment when the penny drops on just what is gravity. Whether for dinner party conversation or an educational resource for students and teachers alike, we hope you'll find this a collection worth talking about.

Explainers are in the DNA of The Conversation: an ambition to unlock the knowledge within the academic and research sector, and deliver it direct to the public.

We're delighted to be collaborating with CSIRO on this book and for their support and encouragement as a Founding Partner of The Conversation.

Andrew Jaspan
Editor, The Conversation

FOREWORD 2

The book you are holding in your hand is an offshoot of a phenomenal, innovative news outlet. It will be ideal for taking down to the beach, or on a camping trip, or giving as a gift for sharpening the critical faculties of a young friend or relative. It represents a selection of articles from two important sections of The Conversation: 'The Explainer' (in-depth articles on issues of the moment) and 'Medical Myths'.

If you are not yet familiar with The Conversation, you may not realise how ground-breaking it is. During the 2010 Federal election campaign, I remember sitting down at breakfast to read some election coverage in one of our national broadsheets. I was interested in an analysis of economic and environmental policy, tax reform and immigration. What I got, for the most part, was commentary on the leading candidates' dress sense and personal styles. I remember thinking, 'I just don't want to do this anymore', and have rarely bought a paper since. Instead, I have increasingly drawn on a combination of weekly magazines for analysis, radio for up-to-the-minute news, and, increasingly, The Conversation for deep insight. It is a unique combination of a news outlet and a scholarly journal. It covers topical issues, but the person writing will have actually established a track record as a specialist on the topic. Instead of say, reading the thoughts of a journalist on school funding reforms whose experience as an educational commentator is limited to an early career reporting on school sports days, you get a reasoned analysis, with graphs, prepared by someone with a PhD in education policy. There will be links in the argument to the research literature, so you can read more deeply on the subject, if you are interested. While the writing will be less racy than typical newspaper journalism, it will be accessible to non-experts, because of the hard work on readability done by the editorial staff.

It seems that I am not alone in wanting a more intelligent, less personality-based, school playground approach to our national discourse. Since the launch of The Conversation in March 2011, it has become Australia's largest independent news and commentary site, with over 1 million unique visitors a month, and rising. Not only is it having a direct impact on its subscribers, it is helping to raise the level of debate more broadly, as the more traditional mass media outlets are coming to it as a source of experts and ideas, with 69% of The Conversation's articles being republished by others, such as SBS, the ABC and the *Sydney Morning Herald*. The readership is becoming increasingly international, with 35% of the visitors coming from overseas, especially the US and the UK. Indeed, a branch of The Conversation has now opened in the UK.

A former Prime Minister referred to some issues as being 'barbecue stoppers', defined broadly as a controversial issue likely to interrupt a barbecue by provoking vigorous debate. I do think at times that we in Australia could benefit from a more combative conversational culture at social events. Armed with this book, perhaps we can start a movement for real conversation in Australia! Before a barbecue, you can bone up on such starter topics as:

'Does luck exist?'

'Do deodorants cause breast cancer?'

'What is stuttering?'

'Why is the sky blue?'

'What are quarks?'

You can do more than simply start the discussion – with this book you can marshal the explanatory information to help inform it. Show some restraint, though: there is a fine line between a well-informed barbecue guest and a bore. I note also that a secondary meaning of barbecue stopper is 'a social gaffe'. A good indication of whether you are striking the right note will be follow-up invitations.

The Conversation is playing an increasingly important part in our national discourse, used by many journalists as a source of insight into the news behind the news. In an era of bizarre Internet conspiracy theories, the screeching of online forums and inconsequential twittering, The Conversation is like a walk beside a deep, calmly flowing river of knowledge. As Thomas Jefferson said, 'A well-informed populace is vital to the operation of a democracy'. It is a privilege reading the thoughts and insights of people who really know their subject. I commend this book to you.

Dr. Mark Lonsdale
CSIRO
Member of the Editorial Board, The Conversation
Member of the Advisory Committee, CSIRO Publishing

ACKNOWLEDGEMENTS

The Publisher wishes to thank the authors for their contributions, The Conversation for their support of this project and Virginia Tressider for her editorial input.

CONTENTS

'Nothing is so dangerous to the progress of the human mind than to assume that our views of science are ultimate, that there are no mysteries in nature, that our triumphs are complete and that there are no new worlds to conquer.'

Sir Humphry Davy

'Science, my lad, is made up of mistakes, but they are mistakes which it is useful to make, because they lead little by little to the truth.'

Jules Verne, *Journey to the Centre of the Earth*

'Nothing in life is to be feared, it is only to be understood. Now is the time to understand more, so that we may fear less.'

Marie Curie

'Nature composes some of her loveliest poems for the microscope and the telescope.'

Theodore Roszak, *Where the Wasteland Ends: Politics and Transcendence in Post-Industrial Society*

Animals and agriculture

'We have become, by the power of a glorious evolutionary accident called intelligence, the stewards of life's continuity on earth. We did not ask for this role, but we cannot abjure it. We may not be suited to it, but here we are.'

Stephen Jay Gould, *The Flamingo's Smile: Reflections in Natural History*

'No matter how rhapsodic one waxes about the process of wresting edible plants and tamed animals from the sprawling vagaries of nature, there's a timeless, unwavering truth espoused by those who worked the land for ages: no matter how responsible agriculture is, it is essentially about achieving the lesser of evils. To work the land is to change the land, to shape it to benefit one species over another, and thus necessarily to tame what is wild. Our task should be to delivery our blows gently.'

James E. McWilliams, *A Revolution in Eating: How the Quest for Food Shaped America*

'Agriculture not only gives riches to a nation, but the only riches she can call her own.'

Samuel Johnson

'I know of no pursuit in which more real and important services can be rendered to any country than by improving its agriculture, its breed of useful animals, and other branches of a husbandman's cares.'

George Washington

How are vaccines used in Australian agriculture?

Ian Colditz, Research Scientist, Animal, Food and Health Sciences, CSIRO

Approximately 140 vaccines are registered for use in livestock and companion animals in Australia. Many more animals than humans are vaccinated each year.

Vaccines are used in farm animals:

- to protect livestock against endemic diseases;
- to modify reproductive performance (for instance, by preventing sexual maturity in young males);
- to improve food quality (for instance, to reduce boar taint – the unpleasant smell from the meat of some male adult pigs);
- to reduce the risk of transmission of diseases such as Hendra virus from horses to humans;
- to produce diagnostics products for use in pathology services; and
- to produce therapeutic products for use in human and veterinary medicine.

Most decisions to vaccinate farm animals are made by livestock owners on a commercial basis. They balance the cost of vaccination against the risks of disease, reduced growth rates and compromised animal welfare.

An important benefit of vaccination – both for the farmer and more broadly for the community – is reduced reliance on antibiotics for treating infections in farm animals.

Adaptive immunity: learning from the environment

All animals are subjected to attack by microbes and parasites. In return, animals have well-developed molecular and cellular defence mechanisms to fight off and kill infectious agents.

Within the time span of each animal's life, it undergoes non-genetic (phenotypic) adaption to its local environment. Living in the environment leads to changes in physiology, behaviour and immune functions that enable the animal to fine-tune its ability to cope and thrive.

Environmental conditions are learnt and remembered by the physiological, behavioural and immune systems of the animal. For the immune system, the lessons learnt from infection by a disease agent are remembered primarily by lymphocytes (a type of white blood cell) and are recalled when the animal is again exposed to the same disease-causing agent.

The recalled immune response is faster and more effective at clearing the infection. The lessons learnt from some infections such as orf virus (scabby mouth) in sheep are usually remembered for life, with a single infection inducing lifelong immunity to re-infection by the same disease agent. In contrast, some infections induce no effective immunity. In other instances immunity can wane over a matter of months.

Vaccines aim to induce protective immunity by controlled exposure to fragments of disease-causing organisms without exposure to the disease itself.

Passive immunity: animal vaccines helping humans

Offspring receive a cultural inheritance of acquired knowledge about local disease threats from their mothers in the form of anti-bodies. Depending on the species, these are acquired via the placenta, egg yolk, colostrum (a precursor to milk) or milk itself. Maternal antibodies provide passive immunity to offspring for the first few weeks of post-natal life. Some vaccines can be used during pregnancy in animals to enhance antibody transfer to offspring.

Antibodies from animals can also protect humans. Indeed, the first Nobel Prize in Physiology or Medicine was awarded to Emil von Behring in 1901 for his development of serum therapy. Von Behring used blood serum from sheep and horses immunised with *Corynebacterium diphtheriae* to treat patients suffering from diphtheria.

Following von Behring's example, the use of antisera raised in animals to treat humans for systemic diseases such as tetanus has been commonplace for many decades. In Australia, horses continue to be vaccinated to generate anti-toxins to tetanus, snake venoms and other toxins. Australian sheep are vaccinated to produce antivenin against rattlesnake venom for use in America.

In the 1970s, it was found that oral ingestion of antibodies isolated from the colostrum of cows immunised with human gut pathogens can protect humans from a range of gut infections. Products containing antibodies isolated from colostrum of immunised cows protect humans from rotavirus infections, traveller's diarrhoea and dental caries. Similar products are also used in animals.

As the efficacy of antibiotics for control of bacterial infections has diminished, there has been a resurgence of interest in 'passive immunisation'. This uses antisera produced in animals for prevention and treatment of disease in humans and farm animals.

For instance, the prevalence and severity of diarrhoeal disease in humans can be reduced by daily ingestion of colostrum-based products from ruminants immunised with the disease-causing agent (and possibly also by consumption of fresh unpasteurised milk from the same animals). There is a very large potential to implement this technology in developing countries to help control diarrhoeal diseases.

Can vaccination and alternative farming mix?

Vaccination is usually used as part of an integrated disease control strategy in animals. Eradication and quarantine are the most effective strategies; however, eradication is rarely achievable. Vaccination played an important role in eradication of equine influenza from Australia in 2008. Indeed Australia is the only country to have eradicated this disease. Selecting breeding stock for resistance to disease is also a very important disease control strategy in farm animals.

As with all foods, medicines and therapies used in humans or animals, there are divergent views on the merits of using vaccines in animals. Some agricultural production philosophies, such as organic farming, discourage use of vaccines.

However, if regulatory authorities require vaccination, there is an adverse disease history, or if a veterinarian recommends it, the organic certification bodies Australian Organic and Organic Growers can authorise the use of vaccines to aid in disease control on organic farms.

This approach provides an argument neither against organic farming nor against vaccination. A diversity of farming practices and production philosophies is likely to strengthen food security in the face of changing environmental threats and consumer preferences.

What do we know about why whales strand themselves?

Mark Hindell, *Professor, Institute for Marine and Antarctic Studies, University of Tasmania*

Whales are a highly specialised group of mammals which left their terrestrial ancestors for the ocean about 50 million years ago. They have become so well adapted to the marine environment that they can no longer survive for long if they find themselves stranded on a beach.

By and large, they are able to avoid getting themselves in that situation but sometimes, something goes wrong and they become stranded. These events are thankfully quite rare, but they inevitably generate considerable interest.

The first question everyone asks is: why have the animals become stranded in the first place? As is so often the case in biology, there is no simple answer.

In recent years some strandings overseas have been linked to the operation of high intensity sonar. However, the cases where this link has been unambiguously established are relatively few, and have involved a group of whales known as beaked whales, which are deep-diving species seemingly susceptible to sonar. These types of human activities have never been linked to a stranding in Australia.

There are probably as many underlying causes as there are actual strandings. However, by patiently compiling data from each new stranding we are able to tease out some common threads.

For example, we know that some beaches have far more strandings than would be expected if they were completely chance events. The beach at Strahan on the west coast of Tasmania is a good example of a stranding 'hot-spot', with several mass strandings there in the last decade.

In these cases, it is likely that the beach morphology, or shape, plays an important role. Offshore sand banks can create a gutter parallel to the beach. Whales that have moved into shallow water, perhaps in pursuit of prey, can become stuck as the tides recede.

There is also a temporal pattern in the occurrence of strandings: some years have many more strandings than others. The first recorded stranding in Australia was in Victoria in 1825. There is a distinct increasing trend in events since that time, but this is almost certainly a result of the spread of the population around the coast and the increasing time that Australians have been spending on beaches, rather than a reflection of an increasing number of actual strandings.

Nonetheless, after taking this into account there is still a clear 10–12 year cycle in the number of animals stranding. This is related to a climate feature known as the zonal (westerly) and meridional (southerly) winds.

Persistent zonal and meridional winds result in colder, nutrient-rich waters being driven from the sub-Antarctic to southern Australia, resulting in increased biological activity in the water column during the spring months. This has the effect of attracting whales closer to shore than in other years, making them more vulnerable to stranding (but obviously not causing it).

At any mass stranding, the immediate focus of the local wildlife authorities is understandably the rescue and release of any surviving whales. This is an extraordinarily difficult task: the animals can weigh over 40 tonnes, be partially buried in sand and be lying in a vigorously active surf zone.

In recent years the Tasmanian wildlife authorities (perhaps by virtue of the fact that Tasmania has more mass-strandings than any other state in Australia) have developed techniques that have seen them successfully re-float and release a number of sperm whales and pilot whales that would certainly have otherwise died.

The fate of these animals remains an important but unanswered question. There is the possibility that the animals have suffered internal injuries from lying on the beach, or that the disruption of the strong social bonds may be a problem for the inevitably small number of animals that survive.

The only way to really answer this question is to follow the animals in the days and weeks following their rescue. While this is impractical, attaching satellite tags to released whales could provide the same information. This has been done on a small scale with released pilot whales in Tasmania, but widespread implementation is dogged by practical difficulties, such as tag damage, either from contact between whales or with the sea floor in shallow areas.

For now rescuers must be content with the knowledge that the animals have not re-stranded in the area as the only indicator of their fate.

What is biodiversity and why does it matter?

Steve Morton, Honorary Fellow, Ecosystem Sciences, CSIRO
Andy Sheppard, Theme Leader, Biodiversity, Ecosystem Sciences, CSIRO
Mark Lonsdale, Chief, Ecosystem Sciences, CSIRO

'Why should I care about biodiversity?' It's a valid question, particularly in a world that faces a changing climate. In addition, there are other things to worry about such as global food shortages, getting the kids to school on time and exercising.

What is biodiversity?
One simple but profound answer is that all of us need to breathe, drink and eat. These are all benefits that are fundamentally provided by biodiversity. But the reasons to pause and consider the value of maintaining our country's biodiversity are broader than this.

First of all, what exactly do we mean by biodiversity? Biodiversity collectively describes the vast array of about 9 million unique living species (including *Homo sapiens*) that inhabit the earth, together with the interactions amongst them.

The concept includes every species of bacterium, virus, plant, fungus and animal, as well as the diversity of genetic material within each species. It also encompasses the diverse ecosystems the species make up and the ongoing evolutionary processes that keep them functioning and adapting.

We can't get by without it
Without these organisms, ecosystems and ecological processes, human societies could not exist. They supply us with oxygen and

clean water. They cycle carbon and fix nutrients. They enable plants to grow and therefore to feed us, keep pest species and diseases in check, help protect against flooding and regulate the climate.

These benefits are known as ecosystem services. A functioning natural world also provides a living for farmers, fishers, timberworkers and tourism operators, to name but a few. So biodiversity keeps us alive, but there are other, less tangible benefits.

Recreation (such as fishing or hiking), the aesthetic beauty of the natural world and our spiritual connection with nature; the cultural values we place on plants and animals such as the kangaroo and emu on the Australian coat of arms – these are all benefits of biodiversity.

Research suggests that natural environments have direct and positive impacts on human well being, despite the highly urbanised modern lifestyles most of us have. Mental health benefits from exercising in natural environments are greater than those gained by exercising in the synthetic environment of the gym. Mood and self-esteem benefits are even greater if water is present.

The value that humans gain from biodiversity reminds us that, despite being predominantly urban, we are still intrinsically part of the natural world. We are a component of the ecosystem and therefore dependent on it. This has led to the global concerns around anthropogenic biodiversity loss.

Biodiversity in decline

Changes in surrounding biodiversity affect all of us. Unlike other species, however, we have the chance to determine what these effects might be. In considering our role in biodiversity, there is some good news and some bad news.

Let's start with the bad. Globally, biodiversity is in rapid decline. The explosion of the human population from 2 to 7 billion in just 100 years has caused the extinction of many species.

Scientists agree that the earth is experiencing its first anthropogenic climate-driven global extinction event. They also agree that this is happening at a rate too fast for species to adapt. CSIRO research shows that by 2070, the impacts of climate change on Australia's biodiversity will be widespread and extreme. You can find this research at http://www.csiro.au/nationalreservesystem.

This loss of biodiversity is concerning because of the growing consensus that it goes hand-in-hand with a reduction in the stability and productivity of ecosystems. The result may be that the services on which we rely could be compromised in damaging ways.

CSIRO's report *Our Future World 2012* recognises biodiversity decline as one of the megatrends that could severely impact Australia over the coming decades. You can download this report at http://www.csiro.au/en/Portals/Partner/Futures/Our-Future-World-report.aspx.

We have the science: policy is the next step

And the good news? In Australia, we are well placed to meet the challenge of biodiversity management head-on. We have substantial national scientific expertise to draw on. On the global scale we have a good record of effective interaction between science and policy. The latter is particularly important.

To halt the decline in biodiversity across the continent, we must translate accumulated knowledge on biodiversity into government policy. This can be done through programs and on-the-ground management. Tough decisions need to be made about where to invest, what to manage, and which approach to take.

These decisions can be emotionally and politically charged. Navigating the complex environmental, economic and political values can be extremely challenging.

Good resources for good policy

Despite these challenges there are things we can do. Australian scientists are actively developing better ways to support good governance and effective investment for improved conservation decision-making.

- The Environmental Decision hub of the Australian Government's National Environmental Research Program is tackling gaps in environmental decision-making, monitoring and adaptive management. One of the hub's projects assessed approaches to species relocation in Australia (Sheean *et al.* 2012). Relocation is becoming more prevalent as species experience habitat loss due to impacts such as climate change. The scientists have developed guidelines to improve relocation's success rates.

- The Atlas of Living Australia brings together Australia's biological information online, making it quicker and easier to undertake biodiversity assessments (or just look up a species you're interested in). It has 33 million records and is growing by the day. You can find it – and contribute to it – at http://www.ala.org.au/.
- A collaborative project between Indigenous Protected Area (IPA) managers, traditional owners, the Australian Government and CSIRO developed guidelines for IPA management plans. These connect traditional knowledge, law and customs with international systems for protected area management.

We urge you to take a moment to consider biodiversity. Debate about the value of biodiversity (both globally and to you as an individual) will help clarify society's objectives for biodiversity management. It will ensure that the changes we make help to conserve our natural assets for future generations.

Reference

Sheean V, Manning AD, Lindenmayer DB (2012) An assessment of scientific approaches towards species relocations in Australia. *Austral Ecology* **37**, 204–215.

Why is Hendra virus so dangerous?

Peter Daniels, *Assistant Director, Australian Animal Health Laboratory, CSIRO*

The release of a Hendra virus vaccine for horses in late 2012 was seen internationally as a most significant event. This is the first licenced vaccine for use in humans or animals for any of the world's most dangerous human infections, sometimes classified as biosecurity level 4 agents. But what is Hendra virus, how is it spread, and why is it so dangerous?

Hendra is a new emerging infectious disease agent discovered in 1994, when it caused the deaths of 13 horses and a racehorse trainer at a complex in the Brisbane suburb of Hendra.

The first case, a mare, was stabled with other horses. She became ill and died two days later. A further 18 horses later fell ill, and 13 died. The remaining horses were subsequently euthanised to prevent relapsing infection and possible further transmission. The trainer and a stablehand nursed the first case, and both fell ill with an influenza-like illness within a week of the first horse's death. The stablehand recovered, but the trainer died of respiratory and renal failure.

Scientists at the Australian Animal Health Laboratory in Geelong, Victoria, quickly isolated and identified the virus, establishing that it had not been reported anywhere else in the world. Hendra is now recognised as one of the most dangerous virus infections for humans.

What does Hendra virus do?

Hendra virus causes generalised disease in horses and people. Its effects are felt throughout the whole body, but in particular it causes a viral pneumonia or encephalitis. Either of these or the combination can be fatal, and four of the seven persons known to have been infected died. There have been more than 35 outbreaks in horses since the virus emerged, and it has resulted in the deaths of more than 80 of these horses in Queensland and northern New South Wales.

How is it spread?

The seven people who have been infected with Hendra virus all contracted the disease while handling infected horses. The concern is that people could be exposed and infected before there are any suspicions that horses might have Hendra virus. That possibility is to be treated with great concern.

Person to person transmission of Hendra virus hasn't been recorded, but the number of human cases is low. We don't believe humans can spread the virus among themselves but it may not be impossible.

There is one record of Hendra virus infection in a dog. In 2011 a dog living on an infected property was reported to have HeV antibodies, the first time an animal other than a flying fox, horse or human has tested positive outside an experimental situation.

Fortunately, Hendra outbreaks are relatively rare. The virus doesn't spread easily among horses. There may have been cases where this has occurred but it's usually been spread only to one or two animals in the immediate vicinity. It is always hard to know if the horses in the vicinity of a case were infected from the horse or from the primary source of infection.

And that primary source of infection is fruit bats. Hendra virus infection is quite common among them but doesn't seem to cause disease. Problems arise only occasionally where the infection is transferred to horses.

We can't say with absolute scientific certainty how the infection is transmitted from bats to horses. The possibilities include that the virus might be present in the urine of flying foxes or in the saliva of the flying foxes or it might be that when the flying fox gives birth the virus is excreted.

It would seem horses become infected by eating contaminated material, so eating grass contaminated with any of those fluids from flying foxes could be a source of infection. The virus amplifies within the horse, and humans who are exposed to a large amount of the secretions or blood from an infected horse can become infected.

Laboratory studies show that horses may excrete the virus for up to 72 hours before showing clinical signs.

What does the Hendra virus vaccine do?

The vaccine is given to horses, as a way of protecting both them and the people who handle them. Scientific studies have shown that vaccination greatly reduces the amount of virus present in and excreted by horses that subsequently become infected. Because all human infections with Hendra virus have been from exposure to infected horses and direct contact with their bodily fluids, vaccinating horses breaks the chain of virus transmission from flying foxes to horses, and then to people.

The Equivac® HeV vaccine is an important step towards breaking the transmission cycle of this disease, and reducing its impact on the horse-owning community. But it's important to ensure that we continue to protect the health of our animals and people. And to do this, we need to maintain and continue undertaking research and adding to the tools in our armoury of weapons against the deadly Hendra virus.

Body

'Our bodies are apt to be our autobiographies.'

Frank Gelett Burgess

'The human body is not a thing or substance, given, but a continuous creation. The human body is an energy system which is never a complete structure; never static; is in perpetual inner self-construction and self-destruction; we destroy in order to make it new.'

Norman O. Brown, Life Against Death: The Psychoanalytical Meaning of History

'[Think] of an experience from your childhood. Something you remember clearly, something you can see, feel, maybe even smell, as if you were really there. After all, you really were there at the time, weren't you? How else would you remember it? But here is the bombshell: you weren't there. Not a single atom that is in your body today was there when that event took place . . . Matter flows from place to place and momentarily comes together to be you. Whatever you are, therefore, you are not the stuff of which you are made. If that doesn't make the hair stand up on the back of your neck, read it again until it does, because it is important.'

Steve Grand, *Creation: Life and How to Make It*

'We are biology. We are reminded of this at the beginning and the end, at birth and at death. In between we do what we can to forget.'

Mary Roach, *Stiff: The Curious Lives of Human Cadavers*

Can you pay off your 'sleep debt'?

Leon Lack, *Professor of Psychology, Flinders University*
Siobhan Banks, *Senior Research Fellow, Centre for Sleep Research, University of South Australia*

Ever have those moments on weekends or public holidays when you wake at your usual time, then realise there's no pressing need to get up? If you go back for another couple of hours of shut-eye and use the handy excuse of 'paying off your sleep debt', you're not alone. But can we really catch up on lost sleep?

Before we get to the answer, let's look at the two main determiners of sleep and wakefulness: circadian rhythms and the homeostatic sleep drive.

Your circadian rhythm is often described as the body's 'natural pacemaker'. It controls a range of bodily cycles including the 24-hour cycle that regulates your degree of alertness at various times of day. The circadian rhythm effect on sleep continues on a 24-hour schedule, regardless of how much or little sleep we get.

Sleep debt relates to the second determiner: the homeostatic sleep drive or sleep pressure. This drive operates on a simple deprivation and satisfaction model – the longer you go without sleep, the sleepier you get as the sleep pressure or debt builds up.

Sleeping then reduces sleep pressure or 'pays off' the debt. This process can be illustrated simply, as shown in Fig. 1.

When we wake up after a good night's sleep and stay awake across the day, sleep pressure increases – rapidly at first then more slowly. Then after falling asleep, sleep pressure decreases, as the solid non-linear curve shows.

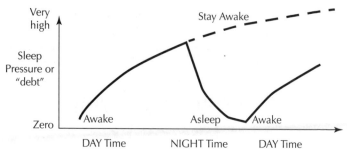

Fig. 1. Homeostatic sleep debt.

The diagram also demonstrates that not all sleep is equal in its ability to pay off accumulated sleep debt. The first few hours of sleep do this more quickly and efficiently that the last hours.

Feeling sleepy?

About a quarter of the Australian population reports rarely getting an adequate amount of sleep. Many more who feel all right would also probably get more sleep and feel better if they allowed themselves more opportunity to sleep.

Consistently getting too little sleep is a particular issue for new parents, who may be woken up every couple of hours to tend a hungry infant.

Shorter sleeps result in a higher starting point of sleep pressure. So for the chronically sleep-restricted, the curve in the diagram above would start from a higher sleepiness value and increase in the same way across the day, but at a higher level.

This sleepiness gradually increases across days and there is strong experimental evidence that this may result in higher rates of functional impairment.

Back to the weekend sleep-in – this allows us to catch up on that lost sleep, doesn't it?

It seems to; however, the experimental evidence for this is pretty sparse and not yet clear. What is clear from some studies is that sleeping in very late can delay the circadian rhythms and make it difficult to get satisfactory sleep the following week. So extending sleep on the weekends might not be that helpful after all. It's better to get consistently adequate sleep across the week than trying to catch up on lost sleep at the weekend.

But what happens if you completely deprive yourself of sleep and stay awake instead of sleeping at night?

You guessed it – your sleep debt continues to increase. This is illustrated in Fig. 1 by the dashed line.

Many total sleep deprivation studies have shown that you can fully recover after one to two nights of longer-than-normal sleep. Following a single night without any sleep, the recovery sleep the following night needs to be only two to three hours longer than normal to return most functions and mood to normal. After four or five nights of total sleep loss, a couple of longer sleeps are usually all that's needed (12 hours, then 10 hours, for example).

Following the documented record sleep loss of 11 days by Randy Gardner, at Stanford University's Sleep Research Center, an extra seven hours on the first recovery night sleep and extra three hours sleep on the second night seemed to return the 17-year-old to normal alertness and functioning, without any long-term negative consequences.

So a 'debt' of 88 hours of sleep loss seemed to be repaid by an additional 10 hours over Randy's normal sleep need of eight hours.

In short, evidence suggests that lost sleep *can* be recovered but our sleep debt doesn't need to be repaid hour for hour.

Do we need to follow medication use-by dates?

Lisa Nissen, *Professor (Head), School of Clinical Sciences,*
Queensland University of Technology

It's late at night. And after a long day at work, you have a splitting headache. You rattle around in the bottom drawer of the bathroom vanity to find a packet of paracetamol tablets you know are hiding there.

Phew, relief is at hand! Then you turn the packet over and discover that the crumpled box of pills actually expired two years ago.

So are they really out-of-date, or would it be okay to take them just this once?

Well, the answer varies, depending on the type of medication, how it has been stored, over what period of time and whether it's still in its original, sealed container.

Expiry dates were introduced for medicines around 30 years ago to signal the date beyond which the manufacturer can't guarantee the full effectiveness and safety of the medication in its original sealed packaging. But that doesn't mean it will suddenly become completely useless the day after it expires.

Once the container or seal is opened, the expiration date on the packaging doesn't apply any more. Just as with a tin of tuna from the cupboard, you couldn't expect it would still be okay in two years time once you opened the tin.

Some tablets and capsules are packaged in blister packs or foil wrappers, so although the box may be opened, the tablet or capsule is still in the sealed container and the expiry still applies. But when the medication comes in an open bottle, once it's opened and exposed to air, the expiry date on the packaging no longer applies.

The United States Department of Defense's Food and Drug Administration Shelf Life Extension Program has investigated the extent to which products are stable after their expiration date has been reached. It found that when medications were stored in their unopened containers, most remained stable for several years past the labelled expiry date.

Over time, the potency of some of the medications declined, while others were open to microbial contamination if there product's preservative became ineffective. Liquid medications such as solutions and suspensions were generally not as stable as tablets and capsules.

Some medications – such as glyceryl trinitrate tablets (Anginine), insulin, eye drops and antibiotics – have specified discard dates that kick in after they're opened. The stability of these medications is significantly altered with exposure to light, heat, moisture and oxygen, causing them to lose effectiveness. These are medications which should never be used after their expiry date.

And finally, where you keep your medication is also very important. For most medications, storing them in a cool place away from heat and high humidity can help to improve their shelf-life.

So what does all this mean to our headache sufferer with his packet of expired paracetamol?

The medication is still in its original sealed blister pack even though it's two years past its expiry date. But it's been stored in a warm and humid bathroom environment, which may alter the stability of medication and could change the effectiveness of the product. If he has no other alternative than the expired paracetamol, it may be better than nothing but it probably won't be 100 per cent effective.

But, of course, the situation would be different if we were talking about insulin, antibiotics or medicines with a 'discard date'. These medicines should never be used beyond this date.

As a rule, it's best not to use products after their expiry date to ensure you're getting the maximum benefit from your treatments. And if you're in doubt about the safety or effectiveness of any medication, check with your pharmacist.

Overweight, obese, BMI – what does it all mean?

Tim Crowe, *Associate Professor in Nutrition, Deakin University*

Australians are getting fatter and there's no dispute over how this increasing weight is affecting our health. Different methods of assessing body fat can give different interpretations of just how much excess weight a person is carrying, but all methods point in the same direction when applied over time.

The most common measure of body fat and associated health risks is body mass index (BMI). BMI was developed as a simple way to compare different groups of people, based on the correlation between height and weight as an indicator of excess body fat.

BMI is calculated by dividing weight in kilograms by the square of height in metres. A healthy BMI for an adult is between 18.5 and 25 kg/m^2. Between 25 and 30 kg/m^2 is considered overweight and 30 kg/m^2 and above obese. At a population level, high rates of body fat above 30 kg/m^2 indicate an increased risk of death and disease. You can find a simple online BMI calculator at http://www.betterhealth.vic.gov.au/bhcv2/bhcsite.nsf/pages/bmi.

BMI is the main measure used for international obesity guidelines and is recommended by the World Health Organization. But BMI isn't perfect. People with the same bodyweight and height can have different proportions of body fat to lean muscle mass. Athletes with high muscle mass, for instance, can have a lower proportion of body fat than less muscular people, so a BMI calculation can put them into an overweight or obese category, even though their risk of obesity-related disease is very low.

This is a frequently cited criticism of BMI, but it needs to be put into perspective. Such people are in the minority and a quick visual inspection will clearly show that it's muscle, not fat, that they're carrying.

People from an Asian background tend to have more body fat on a leaner frame, so a lower BMI healthy-weight-range can be used. Conversely, people from Pacific Islander backgrounds tend to have a higher proportion of lean muscle mass compared with fat, so a higher BMI healthy-weight-range is often recommended.

While BMI is a useful measure of overall health risks, it fails to take into account the distribution of fat throughout the body. For this reason, waist circumference was developed as a simpler and potentially more accurate measure of disease risk. Waist circumference is not only a gauge of body fat, but it specifically targets the most dangerous type of fat: visceral fat.

Visceral fat is found between the organs of the abdomen and contributes to belly fat. There's a strong correlation between central obesity and cardiovascular disease, insulin resistance, type 2 diabetes, inflammatory diseases, high blood pressure and other obesity-related diseases.

For men, the aim is to have a waist circumference below 94 cm; for women it's 80 cm. Measures above 102 cm for men and 88 cm for women carry a very high risk of developing type 2 diabetes, high blood pressure, cardiovascular disease and even some forms of cancer. For people of an Asian background, slightly lower waist circumference goals apply: under 90 cm for men and 80 cm for women.

More recently, estimates of body fat percentage and health risks have looked at waist-to-hip ratio and even waist-to-height ratio. Both these measures take into account central fat stores so can give a better health risk estimate than BMI. For men, a waist-to-hip ratio below 0.9 and 0.8 for women correspond to a healthy weight BMI.

An advantage of using waist measures for body fat estimates is that it takes away the stigma of needing to step on the scales. It also allows for the use of cut-off values that avoid terms of overweight and obesity, and instead focus on the risk of metabolic disease – disruption of the biochemical processes involved in the body's normal functioning.

Labelling a person 'obese' may not always be helpful in bringing about positive behaviour changes, especially when a person already acknowledges that they're carrying a bit of extra weight. 'Unhealthy BMI', 'above the healthy weight range', and 'excess weight' can all carry the same message about the need to shed excess weight for better health and reduced risk of disease.

Another technique to measure body fat is by bioelectrical impedance. This method involves passing a small electrical current through the body, normally by a special set of scales that a person stands on. The scales measure water volume and, by the use of special algorithms, arrive at a body fat percentage estimate. The accuracy of these machines can vary dramatically, especially around the lower end of the price range.

By far the most accurate way to measure body fat is by magnetic resonance imaging (MRI), computed tomography (CT) or X-ray scanning, but such methods are not realistic for the public to use and belong firmly in the world of research.

So long as the limitations of a weight assessment method are understood, methods such as BMI and waist circumference are quick and simple validated ways to assess weight and disease risk that can be used by health professionals and the public alike.

What are antibiotics?

*Matthew Cooper, Professor, Institute for Molecular Bioscience,
University of Queensland*
Mark Blaskovich, Senior Research Officer, University of Queensland

You've got a high fever, severe cough and you're finding it hard to breathe – all symptoms of severe pneumonia. Before antibiotics, you'd likely be dead within 10 days. In fact, before we had these amazing drugs, infections caused by bacteria were the leading cause of death.

Today, take a course of antibiotics and most of the time you'll be completely fine. But just what is an antibiotic, and how does it work? And what's all this talk about antibiotic resistance and superbugs?

An antibiotic is a compound that cures infections by killing or slowing the growth of bacteria. The word literally means 'anti-life', from the Greek 'bios' meaning life. The original antibiotics were made by bacteria or fungi to prevent growth or kill other bugs, in what was effectively chemical warfare between species during the colonisation of our planet by microorganisms billions of years ago.

Antibiotic drugs

Thousands of antibiotics have been discovered over the years, but most have only been shown to kill bacteria in the laboratory. Very few have become drugs, because to do that, an antibiotic must be selective – it must kill the bacteria while not harming the cells of the human body.

An antibiotic must also last long enough in the body to get to the site of the infection and kill the bacteria quickly. The first antibiotic, a blue-green pigment extract from a bacterium called *Pseudomonas aeruginosa*, was a compound called pyocyanase. It killed cholera, typhoid, diphtheria and anthrax bacteria, and was used to treat infections in the 1890s. But it was quite toxic, worked sporadically and not equally well on all patients.

Most antibiotic drugs today are variations of substances that were originally isolated from bacteria or fungi, such as penicillin, which was isolated from a fungal mould in 1928, or vancomycin, which came from soil-dwelling bacteria collected in Borneo in 1953.

Some commercial antibiotics used today are still isolated from bacteria grown in large fermentation vats, while others are modified chemically by scientists to improve their activity or reduce the extent of side effects. Totally synthetic antibiotics produced by chemists have been developed in the last 40 years.

While antibiotics were 'discovered' around 100 years ago, they are actually ancient – microorganisms have been producing them for billions of years. Some antibiotics were produced as defence mechanisms against other bacteria, while others started off as messenger molecules between bacteria, then evolved into killing agents to allow different species of bacteria to out-compete others.

How antibiotics work

Antibiotics work by attacking the parts that a bacterium needs to grow, survive and replicate. Several antibiotics, such as penicillin and vancomycin, inhibit the growth of the outer casing of the bacteria, called the cell wall. Just like the walls of a house, without a strong cell wall, the bacteria collapse.

Bacteria have a very different type of cell wall from those we find in human cells (think a double brick insulated bungalow in Canberra for bacteria versus a timber Queenslander in Brisbane for humans). This is because in people, cells are protected by being surrounded by other cells inside our bodies, while bacteria are exposed in the environment and need stronger cell walls.

In this analogy, antibiotics such as penicillin and vancomycin are designed to attack and destroy bricks, but not touch timber, which means they kill bacteria but don't harm us.

Other antibiotics (such as aminoglycosides, erythromycins and tetracyclines) work by inhibiting protein synthesis, which means the bacteria can't function (there is no kitchen or furniture inside the house). Or they block DNA replication (metronizadole and the quinolones, such as ciprofloxacin), which stops the bacteria from reproducing or replicating (only one house is built in the suburb).

Some of these effects are bacteriostatic – they stop the bacteria growing. This gives our body's immune system time to kick in and clear the infection naturally.

Antibiotic resistance and superbugs

Bacteria develop resistance to antibiotics by rapidly evolving to create ways of neutralising them, actively pumping them out of their cells, or preventing them from entering in the first place. Resistance to antibiotics is now a global problem that is increasing all the time.

Paradoxically, antibiotic resistance is also ancient. Excavations of permafrost in Canada have demonstrated that genes coding for a common type of resistance to vancomycin were present over 30 000 years ago. This is because bacteria developed most of the antibiotics we use today millennia ago, and at the same time other bacteria were evolving ways to defend themselves and resist them.

So antibiotics are a very precious, finite natural resource that we need to take much more care of.

Resistance to antibiotics is a growing problem in treating infections, and is often due to incomplete treatments. Not taking the whole course of antibiotics leads to the survival of a small number of bacteria that can tolerate the drug.

Feeding sub-lethal doses of antibiotics to livestock as growth promoters is another potential source of resistance, because it creates resistant populations of bacteria within the livestock. It also results in low levels of antibiotic contamination in the environment.

Research has shown that this leads to superbugs in animals that can get transferred to humans via our food chain. Bacteria are very promiscuous, and able to quickly swap genes coding for resistance between different species, in a kind of 'bacterial sex'. This has led to the global spread of 'superbugs', bacteria that are resistant to many types of antibiotics, and, in some cases, all antibiotics.

In 2010, bacterial infection killed more people than cancer, according to data from the WHO's Global Health Observatory Database. Despite this enormous human cost, most pharmaceutical companies have left the field of antibiotic drug discovery, primarily for economic reasons. They make more money with drugs people take for a long time, such as cholesterol-lowering drugs, than an antibiotic someone may only need for two weeks.

The train is heading down the tunnel, but we are still walking towards it. We may be re-entering an era where simple infections can once again be death sentences.

What are migraines?

Lyn Griffiths, *Professor of Molecular Genetics; Director, Genomics Research Centre; Director, Griffith Health Institute, Griffith University*

If you, or someone close to you, suffers from migraine, you'll know it is much more than your average headache – migraine is a debilitating disorder that can even affect your sight and speech.

Migraine without aura is the most common type of migraine, accounting for about 70 per cent of all cases. This migraine is often characterised by recurrent headaches, nausea, vomiting, and sensitivity to light and noise.

Migraine with aura is less common, affecting around a third of sufferers. This type of migraine is accompanied by neurological disturbances such as visual and speech impairment and muscular changes which are often experienced shortly before or during the early stages of a migraine.

Visual auras may appear as shimmering lights around objects or at the edges of a person's field of vision; they may give the appearance of wavy images or even cause temporary loss of vision.

Non-visual effects can include weakness, speech or language abnormalities, dizziness, vertigo, and tingling or numbness of the face, tongue, hands or feet.

Migraine affects about 12 per cent of the Western population, with 18 per cent of women, 6 per cent of men and 4 per cent of children suffering from the condition. There are considerable differences in the prevalence of migraine culturally, with Caucasian populations having the highest rates globally.

The condition causes a substantial economic burden. In 1990 the Australian Bureau of Statistics estimated the total cost of migraines to the Australian economy at around $721 million per year, mostly due to loss of productivity at work and reduced occupational effectiveness. Health-care costs associated with migraine are also sizeable, with millions of dollars spent on medical consultation and treatment for migraines each year.

Causes
Sensitivity to particular foods and smells, fluctuating hormonal levels, stress and fatigue can prompt migraines in sufferers, but the underlying causes are believed to be genetic and are poorly understood.

As a geneticist, I've been studying the molecular genetics of migraine for over a decade and have identified several genes (including the MTHFR, oestrogen receptor, Notch 3 and TRESK genes) that increase a person's susceptibility to suffer from migraine. Most migraine sufferers (about 90 per cent) have a close relative – a parent or grandparent – who also suffers from the condition.

Current treatments
There are several pharmaceuticals currently available to treat migraine symptoms, but they are not effective for all sufferers. One of the most common classes of drugs used to relieve migraine symptoms is triptans, which work by affecting the serotonin or 5HT1 neurotransmitter system within the brain. They relieve headaches in 30 to 50 per cent of sufferers, but the headache can reoccur, requiring a second dose. Triptans are available in Australia under the names:

- sumatriptan (Imigran and other brands) tablets, nasal spray and injection;
- zolmitriptan (Zomig) tablets;
- naratriptan (Naramig) tablets;
- rizatriptan (Maxalt) wafers; and
- eletriptan (Relpax) tablets.

Unfortunately triptans can have unpleasant side-effects, including sensations of tingling, heat, heaviness or tightness in the chest and throat, as well as flushing, dizziness, drowsiness, dry mouth and a transient increase in blood pressure. Triptan overuse can also sometimes lead to dependence and subsequent withdrawal syndromes.

Other drugs used to treat migraine symptoms include non-steroidal anti-inflammatory drugs (NSAIDs) – the class of drugs that includes aspirin and ibuprofen, anti-nausea and vomiting drugs, and pain medication. The problem with these drugs is that they treat the symptoms, rather than the cause, of migraine, which generally means that they are less effective.

Many migraine sufferers are concerned about the side-effects of these drugs and are looking for safe and effective alternatives to treat or prevent their condition.

A simple solution

Our research has shown that dietary folate and increased vitamin B levels can reduce the frequency and severity of migraine (Lea *et al.* 2009; Menon *et al.* 2012). We have identified a specific gene that plays a role in causing migraine – in particular, migraine with aura. The gene has a mutation that results in a reduced enzyme level.

In our two clinical trials, results have shown that we can overcome the gene mutation by adding increased levels of the enzyme's 'co-factors' to the diet. The co-factors are simple B group vitamins which enable the enzyme to work better despite the mutation.

We found that vitamin B supplementation reduced migraine disability over a six-month period. Headache frequency and pain severity also decreased significantly. These results are very promising and provide the possibility of a simple but effective targeted treatment for migraine sufferers.

The final phase of this research, in which we are determining the specific required supplement dosages, is currently under way. Based on earlier results, we can expect this relatively inexpensive, non-toxic vitamin therapy to have enormous potential to improve the health and quality of life for many thousands of Australian migraine sufferers.

The results of the final trial will be released in 2014 and we expect the treatment to be available shortly after.

References

Lea R, Colson N, Quinlan S, Macmillan J, Griffiths L (2009) The effects of vitamin supplementation and MTHFR (C677T) genotype on homocysteine-lowering and migraine disability. *Pharmacogenetics and Genomics* **19**, 422–428.

Menon S, Lea RA, Roy B, Hanna M, Wee S, Haupt LM, Oliver C, Griffiths LR (2012) Genotypes of the MTHFR C677T and MTRR A66G genes act independently to reduce migraine disability in response to vitamin supplementation. *Pharmacogenetics and Genomics* **22**, 741–749.

What are trans fats?

Peter Clifton, *Professor of Nutrition, University of South Australia*

Trans fats – they're in our chips, bakery goods, popcorn and cakes. We know we should avoid them, but what exactly are they, and why are they so bad for us?

First, let's take a step back and look at how trans fats fit into the two broad categories of edible fats: saturated and unsaturated.

What are saturated fats?

Saturated fats have a stable chemical composition – they're solid at room temperature and oxidise slowly. Because they're very stable and feel good in the mouth, they're commonly added to processed foods.

Health-wise, saturated fats raise the level of cholesterol in the blood. And in large quantities they can increase your risk of heart attacks and strokes.

Animal fats – such as in cream, butter and milk – tend to be at least half saturated fat. Plant products, such as coconut oil and palm oil, contain saturated fats, as do many prepared foods.

What are unsaturated fats?

The chemical composition of unsaturated fats is much less stable. They're liquid at room temperature, which makes them more difficult to use.

From a health perspective, they actually lower blood cholesterol.

Fats from most oilseeds, avocado and nuts are unsaturated.

What are trans fats?

Trans fats are variants of unsaturated fats, which have been chemically altered to improve their physical characteristics. They're produced industrially to harden fats and oils.

Low levels of trans fats are also found naturally in cow fat and milk.

When trans fats were first introduced to food production over 50 years ago, they were considered miraculous because they allowed a liquid oil to be converted to a solid spread without the adverse effects of saturated fat on blood cholesterol.

The original 1961 US studies of trans fats didn't show any elevation of blood cholesterol, so they were thought to be a healthier option.

What makes trans fats harmful?

Research by Mensink and Katan (1990) showed trans fats elevated the harmful LDL (low-density lipoprotein) cholesterol by about a tenth more than regular unsaturated fat.

And compared with other fats, trans fats didn't have the benefit of elevating the protective HDL (high-density lipoprotein) cholesterol. Mensink and Katan concluded that trans fats were worse for heart disease than the equivalent amount of saturated fat.

This was shown convincingly by Walter Willett and his team in a 1993 study of US nurses (Oh *et al.* 2005). Those who reported eating a large amount of trans fats (more than 5.7 g a day) were around two-thirds more likely to have a heart attack than nurses eating less than 2.4 g a day.

Trans fats from dairy and beef fat ('natural' trans) were not linked to heart disease risk.

How are trans fats regulated internationally?

In 2004, Denmark became the first country to ban industrially produced trans fatty acids at a level of more than 2 per cent of total fat. But it's too early to tell if this has had an effect on heart disease rates.

The United States took a different approach, mandating the labelling of trans fats on food packaging in 2006. This encouraged manufacturers to question the inclusion of trans fats in their food and many removed the product so they didn't have to make this declaration.

The state of New York implemented a partial ban on trans fats in restaurants in 2006, with the ban fully in place in 2008. California and at least a dozen other jurisdictions followed suit, as did Switzerland and Denmark in 2008.

How much trans fats do Australians consume?

Margarines containing trans fats were withdrawn in 1997 and trans fat intake has subsequently declined to a few grams per day.

But even in the 1990s, Australians' intake of trans fat was relatively low, averaging 3.5 g a day in groups at high risk of heart attacks.

We eat a lot more saturated fat than trans fat.

Do you expect Australia will regulate trans fats?

Several high-fat foods – pies, pasties, sausage rolls, quiches, bagels and doughnuts – contain more than 4 per cent trans fats.

Although these foods would generally be regarded as unhealthy because of their saturated fat content, it's important consumers have the option of choosing trans fat-free varieties.

At the very least, trans fats should be labelled so consumers can make their own choice.

Banning trans fats probably isn't required because labelling would serve the same purpose, but with less administrative burden.

Trans fats are much more harmful to health than saturated fat and should be avoided as much as possible.

References

Mensink RP, Katan MB (1990) Effect of dietary trans fatty acids on high-density and low-density lipoprotein cholesterol levels in healthy subjects. *The New England Journal of Medicine* **323**, 439–445.

Oh K, Hu FB, Manson JE, Stampfer MJ, Willett WC (2005) Dietary fat intake and risk of coronary heart disease in women: 20 years of follow-up of the nurses' health study. *American Journal of Epidemiology* **161**, 672–679.

What is a gene?

Merlin Crossley, *Dean of Science and Professor of Molecular Biology, University of NSW*

There's a very confusing exchange in Lewis Carroll's *Through the Looking Glass*:

> *'When I use a word,' Humpty Dumpty said, in rather a scornful tone, 'it means just what I choose it to mean— neither more nor less.'*

When people use the word 'gene' it's also important to know what they intend it to mean. The meaning may depend on what it refers to: carrying a gene, expressing a gene, transferring a gene or discussing how many genes we have.

One reason the definition is so confusing is that the term was coined before we really knew what a gene was. And the effects of genes – inherited characteristics – were observed before we understood genes.

As our knowledge has advanced, the definition of 'gene' has evolved. The information from the ENCODE project, to identify all functional elements in the human genome sequence, will mean the definition needs to be updated again.

The molecular basis of inheritance

Gregor Mendel carried out the first genetics experiments in the 1860s and showed characteristics were inherited.

We have always known that pea seeds grow into pea plants, not kangaroos. What's more, plants with red flowers usually have offspring with red flowers. Children resemble their parents.

Mendel demonstrated that crossing a red pea with a white one could produce peas that were white or red, but not pink.

We sometimes miss this point because we all have features from our two parents, and many features seem to blend. But Mendel showed that some distinct characteristics could be inherited intact. We can think of these as being encoded by a gene.

But neither Mendel nor Darwin used the word. It was first used in 1909 by Wilhelm Johannsen to refer to 'determiners which are present [in the gametes] ... [by which] many characteristics of the organism are specified'.

In 1915 Thomas Morgan found some genes tend to be co-inherited. For example, flies are more likely to co-inherit short wings and red eyes together from one parent, rather than short wings and short legs.

He deduced this might mean certain genes were close together, much like beads on a string. The idea that the genetic material was linear was born. But we still didn't know what it was.

In the 1940s Oswald Avery showed that an enzyme that chews up DNA, DNase, could destroy genes.

We finally knew the genetic material was DNA.

In 1953 James Watson and Francis Crick, along with Maurice Wilkins, and using data from Rosalind Franklin, showed DNA took the form of a double helix. That it was double, with two matching strands, suggested how it could be replicated.

First definitions of a gene

But what precisely was a gene? Crick explained how DNA could be 'transcribed' into RNA (ribonucleic acid) and RNA could be 'translated' into protein. Think of a protein as a biological tool that does something – e.g. the haemoglobin that carries oxygen in your blood.

This gave us our first solid definition:

A gene is a stretch of DNA that encodes a piece of RNA that encodes a chain of protein.

The technical details are complex but let's imagine how you might make a metal axe, or many axes.

Picture a segment of DNA bundled in the precise shape of an axe-head. Imagine the RNA nestling in and forming the impression of an axe-head – just like a mould.

The RNA travels out of the DNA storage room – the nucleus – and you pour in molten iron. It hardens and out comes an axe-head. You would have another mould for the metal handle.

The axe-head and handle bounce around in the cell, find each other and self-assemble. Post-translational modifications, akin to sharpening, can be done by other machines in the cell.

If we mould a lot of axes we say the gene is expressed at high levels. This means that a large amount of its genetic information is present in the final gene product. If there are few or no axes, the gene is expressed at a low level, or is silent.

We can use the axe analogy in another way. One definition of a gene is a region that makes a protein tool. But many DNA genes simply make RNA and the process stops there.

Similarly, the RNA for an axe-head – or one like it – might make a perfect holder for an axe: it doesn't need to go on to make the axe itself.

This gene would produce what is called a 'non-coding RNA' – one that has a function in itself and doesn't need to encode a protein.

Early life forms probably used RNA only, with no proteins or DNA. Our oldest cellular tools – tRNA (which forms the physical link between DNA and RNA and the amino acid sequence of proteins) and rRNA (essential for protein synthesis in all living organisms) – work on the assembly line making proteins but are themselves never translated into protein.

Recent work suggests we have underestimated the number of non-coding RNAs. There also appears to be 'noise' RNA. This probably does little harm but no good either: not all RNA is functional. A segment of DNA that encodes an RNA is not necessarily a gene.

Instead of casts or molten iron, DNA consists of strings of Lego-like blocks of different shapes. A section of the DNA blocks is read into RNA blocks.

The RNA blocks are read into 20 different protein-building blocks that fold according to their shape. The sequence of Lego blocks is dictated by the sequence in the DNA, via a special code – the genetic code.

A definition at last

But now we have a definition for a gene:

> *Genes are stretches of DNA that have the potential to create a tool or a characteristic – such as red colour in the pea flower.*

The outcome is called the 'phenotype'. Our 'genotype' (our genetic material) plus environmental inputs create our phenotype. The Human Genome Nomenclature Organisation defines a gene as 'a DNA segment that contributes to phenotype/function'.

A gene is a linear section of DNA – of a chromosome – that contributes some function to the organism. There are thousands of genes on each human chromosome.

There are also spacer regions between genes and within genes that may or may not do anything.

Some do – the major control region of the gene (the promoter) sits just upstream or around the start point of the gene; while enhancer and silencer elements can be positioned at very great distances along the chromosome, regulating the level of expression.

It is not clear whether to define the control regions as part of the gene. Strictly speaking, the gene is usually only the 'coding part' but mutations in the control regions can be just as damaging as those in the gene itself.

So one good definition of a gene is 'the entire DNA region that is necessary for the synthesis of a functional RNA or protein'.

At first, each gene was thought to produce one protein tool, but we now know that one gene can produce more than one protein or tool.

The gene might be 'spliced' – a process where bits are cut out of the RNA transcript before it is translated into protein. This produces fundamental changes in the organism.

The amount of alternative splicing in humans is extensive. Typically several gene products are made from each gene.

The suggested post-ENCODE definition of a gene is:

A union of genomic sequences encoding a coherent set of potentially overlapping functional products.

Why would we have evolved a gene for cancer?

We can now explain what it means for a plant to carry the gene for red flowers – it may mean that the plant has a stretch of DNA that encodes an enzyme (a protein tool that catalyses chemical reactions) to make a red pigment.

But is there a gene for white flowers? It might just be a mistake in the red flower gene where the enzyme no longer functions, so the flowers have no colour.

This explains the confusion between describing the gene in terms of the tool it makes or the ultimate effect of that tool.

What does carrying the gene for breast cancer mean? It doesn't mean a gene has evolved with the function of causing breast cancer. It means a gene involved in limiting cellular doubling in breast tissue or in DNA maintenance is mutated and no longer functions. So the probability of a cancer growing increases. The gene predisposes the carrier to cancer – it doesn't cause it.

The gene for haemophilia is not there to cause bleeding: a gene has mutated, resulting in a defective clotting factor. Bleeding is the result.

There are several genes for breast cancer and two common genes for haemophilia. Many biological proteins work together or in pathways, and breaking any link in the chain can have serious outcomes.

The most confusing thing is that the 'gene for breast cancer' may have a very indirect relationship to the biology of the breast.

If people were planets and one had a mutation in its axe handle gene, a molecular biologist would observe that there were no functional axes on the planet, but a geneticist would have first noticed that the world was covered in trees.

The gene wouldn't be called the axe gene, but would first be noticed as the gene for making forests. It would only be later that someone would map the gene, clone it and find out what it encoded and how its product functioned.

How many genes do we have?

We still aren't sure how many genes we have. Early estimates were as high as 100 000, but we now think there are many fewer.

One can spot many genes by computer from their key features – an RNA is read from them and the genetic code translates into a protein of reasonable length. But it's hard to identify short genes and functional non-coding RNA genes.

There are probably about 20 000 genes encoding proteins and perhaps as many encoding functional RNAs. We don't know the precise number: it is very hard to be sure which segments of DNA are read into functional products. We won't know that unless they are mutated.

That's an experiment we won't be doing on humans. But as our information on existing human populations and other species increases, we are sure to improve our knowledge of the vast genomic wonderland.

What is cancer?

Darren Saunders, *Cancer biologist, Kinghorn Cancer Centre, Garvan Institute of Medical Research; and Senior Lecturer in Medicine, University of NSW*

Few things strike fear into people more than the word cancer, and with good reason. While improvements in cancer therapy and advances in palliative care mean that the illness does not always lead to inevitable and painful death as it once did, approximately one in three of us will get some form of cancer in our lifetime.

Cancer accounted for about three in 10 deaths (over 42 000) in Australia in 2012. It was the second most common cause of death after cardiovascular disease. Aside from the obvious personal cost, cancer is expensive, with direct costs to our national health system running at $3.8 billion a year.

The US National Cancer Institute defines cancer as a disease in which abnormal cells divide without control and are able to invade other tissues.

Our bodies contain over 200 different types of cell, the basic units of life. Each of these has specific functions and is organised into the various organs such as the lungs, liver, skin and brain. To keep these organs functioning, cells grow and divide to replace other cells as they age and die.

The exquisite balance between cell growth and death is normally kept under tight control by an incredibly complex genetic network. Mutations in the DNA of genes controlling the network can disrupt this balance, causing an accumulation of excess cells, which forms a tumour.

If a tumour forms in an essential organ, such as the liver or lungs, it may eventually grow large enough to compromise organ function and kill us. But some of the most common cancers occur in organs that aren't necessary to keep us alive, such as breasts and prostates. Here, the real problems usually arise when cells from the primary tumour spread (metastasise) to form secondary tumours in essential organs.

It's staggering to think that a mutation in any one of the 100 trillion or so cells in our body is all it takes to initiate a tumour. Cancer-causing mutations can be inherited, or induced by infection with bacteria such as *Helicobactor pylori* – the gastric ulcer bacterium – viruses (human papillomavirus and HIV), or environmental factors such as smoking, asbestos and radiation.

But most mutations probably occur as the result of unrepaired DNA damage that is a consequence of normal cell metabolism.

Complexity is one of the most challenging aspects of understanding cancer and developing therapies. At the molecular level, this complexity can be reduced to a relatively small number of underlying principles known as the hallmarks of cancer. These are:

- sustaining proliferative signalling (keeping the 'grow' switch permanently on);
- evading growth suppressors;
- resisting cell death;
- enabling replicative immortality (normal cells can only replicate a finite number of times);
- inducing angiogenesis (the formation of new blood vessels from existing ones); and
- activating invasion and metastasis.

But the more we learn about cancer at the genetic level, the more we understand that each person's disease is unique.

In other words, cancer is not a single disease. Over 100 different types of cancer have been described using anatomical classifications – that is, by the organ or cell type in which they originate (such as prostate cancer, bowel cancer, breast cancer, skin cancer and lung cancer). But common molecular features (such as a particular genetic signature) are emerging as a much more powerful way to determine appropriate treatment.

This paradigm shift in treatment is being driven by large-scale gene sequencing and functional genomics projects, which are giving us unprecedented insight into cancer at the genetic and molecular level. Australia is at the forefront of this effort, particularly in pancreatic and ovarian cancer genomics.

By far the strongest risk factor for cancer is age. Put simply, the older you get, the better your odds of getting cancer. A long list of genetic and environmental risk factors has been identified for various cancer types, but many of these have relatively moderate effects. Family history is also a predictor of cancer risk but discrete inherited mutations (such as BRCA1 – BReast CAncer Susceptibility Gene 1) have only been linked to a relatively small number of cancers. Obesity is emerging as a common risk factor for several cancers.

Cancer prevention efforts have largely been elusive, with the notable exceptions of the HPV vaccine (for cervical cancer), and reduced smoking rates (for lung cancer).

Until recently, the three pillars of cancer treatment have been surgery, radiation and chemotherapy. The simple aim of all three is to reduce tumour mass.

The first report of cancer being cured by surgery appeared in 1809, and it remained the only effective tool available through the first half of the 20th century. Radical mastectomy for breast cancer – still one of the most profound influences on cancer surgery – was first performed in 1894. Radiation treatment was first shown to cure head and neck cancer in 1928.

Radiotherapy and chemotherapy both work by targeting rapidly dividing cancer cells with radiation or toxic chemicals, respectively. But side effects arise when normal, healthy cells are inadvertently targeted. The cells lining the gut, for instance, and hair follicles are also rapidly dividing and often the victim of collateral damage.

More recent developments in therapy have come in the form of immunotherapy, which uses antibodies or stimulates parts of the immune system to help fight the growth of tumours, and newer therapies designed to specifically target molecular abnormalities related to the hallmarks of cancer. Notable examples of these include Herceptin for the treatment of breast cancer, and Gleevec (Imatinib) for the treatment of some types of leukemia.

There are myriad misconceptions about the risk factors for cancer, and no shortage of snake-oil salesman flogging unproven cures to desperate patients. Everything from coffee enemas to an alkaline diet has been touted as a cure. Cancer Council Australia has a great website that uses hard evidence to dispel many of the myths and misconceptions about cancer. You can find it at http://www.iheard.com.au/.

The best news is that we're starting to beat this disease. And for some cancer types, such as testicular cancer and specific kinds of leukemia, we can effectively treat most cases.

Five-year survival across all types of the disease has increased markedly in the last 20 years (from 47 per cent in 1982–87 to 66 per cent in 2006–10) but these gains have not been consistent across all cancers. Australians generally have better cancer survival prospects compared with people living in other countries. But unfortunately, there are still inequalities for Indigenous Australians, people living in remote areas and people of lower socio-economic status.

Further advances in screening, diagnostics, and targeted therapeutics are likely to continue pushing cancer survival rates even higher. The emergence of personalised medicine has the potential to completely change the way cancer is treated, and there is a rapid push towards this approach, even in the face of the very expensive price tag.

But what of the elusive cure? It's likely that we'll turn cancer into a disease that is managed and treated as a chronic disease, so that people will die with, rather than from, cancer.

What is cerebral palsy?

Nadia Badawi, Macquarie Group Foundation Chair of Cerebral
Palsy, University of Notre Dame, Australia
Iona Novak, Head of Research, Cerebral Palsy Alliance; and
Associate Professor, University of Notre Dame, Australia

Cerebral palsy is the most common physical disability, affecting
35 000 Australians, or one in 500 people. It is estimated that one
Australian child is born with cerebral palsy every 15 hours.

We know cerebral palsy is caused by injury to the developing
brain, which affects the part of the organ responsible for movement. This injury usually occurs before birth, but we often don't
know the cause.

Cerebral palsy is a lifelong condition and can vary in severity
from very mild, where people can walk and climb stairs, through
to extremely severe, which leaves no independent mobility. People
with cerebral palsy usually have a normal life span but experience
the effects of ageing earlier.

How does cerebral palsy affect movement?
Cerebral palsy can affect the body in different ways:

- one in four people with cerebral palsy is unable to walk;
- three in four experience chronic pain;
- half have an intellectual disability; and
- one in three has epilepsy.

In most cases, there is involuntary muscle tightness, known
as spasticity. Spastic quadriplegia involves all four limbs, while
diplegia affects the movement of the legs.

Another common form of cerebral palsy is hemiplegia, the
inability to move either the right or left side of the body, which
can also be caused by strokes.

What are the causes of cerebral palsy?

In the past, it was assumed that most cerebral palsy resulted from a lack of oxygen during birth. But we now know that this is the case for only a small minority.

Most cerebral palsy occurs during foetal development, with extremely premature birth a major risk factor for the condition.

Thankfully, the rate of cerebral palsy has been declining in preterm babies in developed countries such as Australia and The Netherlands, probably as a result of expert neonatal intensive care.

Mothers are now administered steroids before the preterm birth and the premature infants are given caffeine after birth to remind them to breathe during the first few weeks of life.

More than half of all children diagnosed with cerebral palsy were born on time and most were 'healthy' babies. There often isn't an obvious cause in such cases, but there are many known risk factors.

Statistical risk factors include the failure of the baby to thrive during pregnancy, having an abnormal placenta, a family history of cerebral palsy or other neurological conditions.

Other known associations include birth defects, maternal thyroid disease and infections. In the past, severe neonatal jaundice due to problems with the rhesus blood group (the Rh factor) caused cerebral palsy.

But this has been virtually eradicated, due to anti-D (immunoglobulin) therapy – although other causes of jaundice can still be damaging.

Another 10 per cent of cerebral palsy results from a damaging event following birth such as meningitis, road accidents, near-drownings and non-accidental injury.

How is cerebral palsy treated?

There are many effective therapies including casting (holding a muscle in a stretched position with a plaster cast, which helps lengthen the shortened muscles associated with cerebral palsy), muscle strengthening and Botox (which is injected into over-active or spastic muscles to provide a window for therapy to work on strengthening under-active muscles).

A new type of activity-based physical therapy called goal-directed training, which uses regular repetition of meaningful tasks, is also delivering results. We know from research among elite athletes and musicians that the brain forms efficient nerve connections and dedicates more space to tasks that are carried out repetitively and successfully. This appears to be the case in cerebral palsy as well.

Another therapy that appears to help is called constraint-induced movement, which can also be used in infants who have had a stroke. It involves constraining the unaffected arm in a mitt and challenging the weaker arm with appropriate tasks.

Next steps in cerebral palsy research

Recent research has focused on early diagnosis of cerebral palsy to help decrease the rate of complications such as hip dislocation or scoliosis of the spine.

In premature babies, a drug called magnesium sulphate has been shown to decrease the rate of cerebral palsy by 30 per cent in babies of less than 30 weeks gestation. And scientists have just started testing whether magnesium sulphate can also help babies born between 30 and 34 weeks of gestation.

Another exciting recent finding shows that cooling full-term infants who are clearly unwell at birth, and showing signs of neurological abnormality, may decrease the risk of cerebral palsy.

Other protective therapies about to be trialled in humans include melatonin – the hormone that regulates the body's circadian rhythm – and erythropoietin (EPO) – a hormone that stimulates red blood cell production, which have shown promising results in animal trials.

But while hundreds of Australians have sought expensive, risky and unproven stem cell therapy in developing countries, research is still at an early stage in reputable laboratories.

Research in cerebral palsy has historically lagged behind other medical areas. But recent growth in the number and quality of studies offers hope for prevention, better interventions to improve the quality of life of people living with this condition and – hopefully – one day a cure.

What is colour blindness?

Paul Martin, *Professor of Clinical Ophthalmology & Eye Health, Central Clinical School, Save Sight Institute, University of Sydney*

When you look at a rainbow, how many different colours do you see? Most people say seven, but some people would say only two or three. There are even some (very rare) people who see no colour at all.

How can it be that one person says two things have the same colour, yet somebody else says they are completely different?

Colour isn't really there

Scientists know that Isaac Newton did more than just sit around watching apples fall into his garden. He linked gravity on Earth to the movements of planets, and his experiments with glass prisms showed that white light is a mixture of different wavelengths (he called them *refrangible rays*).

One of his many brilliant insights was that unlike size or weight, colour is not a property of the objects that fill our world. Colour depends precisely on which wavelengths of light are bounced from objects before they reach the eye.

Colour is a sensation, a property of the mind. No matter how bright and vivid and real they seem, colours are inside your head, not outside.

Why do people see colours?

Light is picked up in the eye by three types of cells called cone cells. They are called cones because under the microscope they look like tiny ice-cream cones.

They are nicknamed red, green and blue because they pick up different wavelengths. There are millions of each type of cone cell. Just as a painter can mix from three tubs of paint to produce a wide and vivid palette, your brain uses these three cone types to create thousands of colour sensations.

What causes colour blindness?
The cone cells are just like other cells in the body – they are controlled by genes. The genes controlling cones are prone to faulty replication during early development, and affected cones either fail to develop, or start to pick up abnormal wavelengths. The result is like taking away or diluting one of our painter's tubs: the colour sensations are reduced or changed.

How is colour blindness inherited?
Every cell in every woman's body contains two gene packages called X chromosomes, but men have only one. The genes controlling red and green cones are located on X chromosomes.

If a woman has a faulty or missing gene on one X chromosome, the gene on the second X chromosome works as a backup and the cones develop normally.

But if the faulty X chromosome is transmitted from mother to son, there's no backup, and the son will have reduced or altered colour sensations, called *red-green colour blindness*. Other forms of colour blindness are much more rare and usually more severe.

Can colour blindness be cured?
Most scientific studies suggest that the wiring of the eye and brain is identical in people with normal and abnormal colour vision. The only difference is at the first stage of vision, where the cones can be faulty.

The obvious solution is to fix the faulty cones, and this is what a team led by scientists Maureen and Jay Neitz at the University of Washington in Seattle have attempted.

Monkey see, monkey do
Maureen and Jay Neitz study a species of monkey in which all the males are red-green colour blind. In gene therapy trials, their

team injected colour-blind male monkeys with the gene controlling the missing cone type.

We don't know what monkeys see, but we can see what they do. Two monkeys tested so far learnt how to tell red from green patterns after the injections.

These are promising results, but the eye is a very delicate organ and the injections are still dangerous to sight. Colour blindness may be curable but there is still some way to go.

What is deep vein thrombosis?

Karlheinz Peter, *Professor and Lab Head, Atherothrombosis and Vascular Biology, Baker IDI Heart and Diabetes Institute*
Ulrike Flierl, *Research Officer, Baker IDI Heart and Diabetes Institute*

Living in Australia, we're used to flying long distances. So you've probably wondered about the risk of developing a deep vein thrombosis. Perhaps you've even considered buying some pressure stockings for that next long-haul flight?

What is deep vein thrombosis? And what does the evidence say about reducing our risk of developing it?

Deep vein thrombosis is the formation of a blood clot (called a thrombus) in the deep veins of the leg. The clot can be either located in the lower leg or in the thigh, or both. Rarely, a blood clot develops in other veins such as in the arm.

Eventually, the thrombus is in danger of dislocating from the vessels in the leg and going straight into the lung circulation (pulmonary embolism), thus blocking the blood supply of the lung and leading to shortness of breath.

Venous thromboembolism (VTE) – the term which encompasses both deep vein thrombosis and pulmonary embolism – affects around 52 in every 100 000 Australians and is the country's fifth leading cause of death. So early detection and treatment is vital.

On the other end of the spectrum, blood clots can also form in the more superficial veins of the leg, just under the skin. This is called thrombophlebitis and is a much less serious condition.

Who is at risk?

There are three principal mechanisms that increase the likeliness of developing deep vein thrombosis:

- reduced flow of blood (from being immobilised, whether through illness, leg injury, or long sitting during long-haul flights);
- increased tendency to blood clotting (due to hereditary diseases such as Factor V Leiden disease, a mutation of one of the clotting factors in the blood called factor V); and
- injury of blood vessels (from accidents or surgery).

The risk of developing a deep vein thrombosis is increased in patients who have previously had deep vein thrombosis or a pulmonary embolism, and in those with a family history of blood clots.

Other risk factors include cancer (or cancer treatment), taking contraceptive pills containing oestrogen, hormone-replacement therapy, pregnancy and conditions that cause abnormal blood clotting, such as thrombophilia.

Some of these risk factors are modifiable, so there is a chance to reduce your risk by losing excess weight, quitting smoking (as smoking affects blood clotting and circulation), and using contraception methods other than oestrogen-containing pills.

On long-haul flights, car rides or bus trips, exercise your lower calf muscles. Whenever possible, get up and walk around, or raise and lower the heels while keeping the toes on the floor while sitting.

Symptoms

The first signs of deep vein thrombosis are swelling of the entire leg or, more often, one side of the calf. Sometimes there is localised painful tenderness and reddening.

In case of the life-threatening complication of lung embolism, the symptoms are sudden shortness of breath with rapid pulse (heart rate), sweating and coughing up blood. If you have any of these symptoms, see your health practitioner immediately.

After a series of questions about the onset and characteristics of your symptoms and a thorough physical examination, further testing will confirm the diagnosis. The best way to diagnose a suspected deep vein thrombosis is an ultrasound examination of the leg. In the case of a suspected pulmonary embolism, other special imaging diagnostics (computer tomography – CT scan – or scintigraphy, involving radiopharmaceuticals taken internally) need to be applied.

Treatment options

The aims of the treatment are to stop the blood clot from getting bigger, from breaking loose – and drifting into the lung leading to pulmonary embolism – and to reduce the chances of deep vein thrombosis happening again.

Deep vein thrombosis is treated with blood thinners (anticoagulants), usually for a period of three to six months. These are mostly administered as injections in the first days, followed by tablets.

Compression stockings on the lower leg prevent the blood from pooling and subsequent clotting. The stockings should be worn for at least one year and after that whenever immobilised, such as on long-haul flights.

The stockings also prevent one common complication that frequently occurs after deep vein thrombosis: post-thrombotic syndrome, which arises from the damage to the veins caused by the blood clot. The syndrome comprises swelling of the affected leg, pain and skin discolouration.

So, should you have an injection or wear compression stockings when you fly?

Long-haul flights (more than four hours) increase the risk for developing deep vein thrombosis, like every other situation where your movement is restricted. Although few studies have been performed to address this question, the increase in risk seems small.

It's important to assess the thrombosis risk on an individual basis. People at the highest risk of travel-related thrombosis who travel more than three hours at a time should wear compression stockings. The stockings need to be individually adjusted to ensure they don't restrict the blood flow and thereby cause, rather than prevent, thrombosis.

In general, a prophylactic injection of heparin is not recommended and wearing compression stockings on each flight has not been proven to be beneficial. This advice is, of course, different for people who have had a previous venous thromboembolism or who have more than one risk factor for developing blood clots.

In any case, it's important you try to reduce the modifiable risk factors for deep vein thrombosis, particularly when travelling long distances.

What is dengue fever?

Scott Ritchie, *Professorial Research Fellow, James Cook University*

Dengue is caused by four different serotypes (strains) of the dengue virus. People are infected via mosquito bites, and the virus can cause mild to severe illness. Dengue has been spreading through most urbanised areas in the tropics in the last 30 years. Up to 40 per cent of the global population lives in dengue-infected tropics, and an estimated 50 to 100 million cases occur annually.

Classical dengue, the most common type of the illness, is characterised by a high fever, splitting headaches, vomiting, a rash and body aches. It's referred to colloquially as 'breakbone fever'. The severe form of the illness, dengue haemorrhagic fever (DHF), is fortunately rare. Its symptoms include blood plasma leakage, which may lead to shock and, potentially, death.

There's no vaccine or specific medication to prevent dengue but both types can be treated with early diagnosis and fluid-replacement therapy. While fatalities are rare (less than 1 per cent of cases), dengue epidemics can bring illness to thousands of people within weeks, causing chaos in communities and costing millions of dollars.

Dengue in Australia

The illness was common in eastern Australia from the late-19th century through to the mid-20th century, stretching south nearly to Sydney. Large epidemics occurred in eastern New South Wales and Queensland. But when rainwater tanks became less common after World War II with the advent of piped water, dengue-carrying mosquitoes retreated to north Queensland, where ample rains provide year-round breeding sites.

Dengue is now limited to the more densely settled areas of north Queensland, the only region in Australia that has the *Aedes aegypti* mosquito, which can carry the disease. Highly urbanised and feeding almost exclusively on humans, this mosquito loves old unscreened Queensland houses.

And the area is increasingly subject to outbreaks of the illness. Since 2000, there have been 41 dengue outbreaks in north Queensland, resulting in 2524 confirmed cases leading to three deaths – all from the bite of a rather innocuous-looking mosquito.

The increase in dengue activity overseas is also responsible for an increased number of imported cases in Australia. In the dengue-receptive cities of Cairns and Townsville, the number of imported dengue cases has jumped from 10 a year to between 30 and 50 a year in the last four years. All four dengue types have been active in the area, resulting in multiple outbreaks.

Queensland Health staff based in Cairns and Townsville have been able to eliminate the virus in each of the 43 different outbreaks in the region since 1995, preventing the virus from becoming established (endemic) in north Queensland. This is important, because persistent outbreaks of multiple dengue viruses are associated with increased incidence of severe illness and deaths.

Current global dengue trends suggest the virus will become more common in Australia. Rainwater tanks are again common in suburban yards, and outbreaks of the virus have increased overseas, meaning more imported cases of dengue.

Controlling dengue outbreaks

While mosquitoes breed in obvious stagnant water sites such as tyres, buckets, pot plant bases and boats under the mango tree, they also exploit hidden, flooded 'cryptic containers' that are even harder to remove or treat. These include sump pits, telecommunication pits, septic tanks, roof gutters and rainwater tanks.

Dengue cases are also hard to isolate, especially the all-important index case (the initial case imported from overseas that kicks off an outbreak). Australia's love affair with Bali has resulted in hundreds of cases among cash-flushed youth seeking a brief tropical holiday.

All Queensland's recent large outbreaks, from the 500 cases in 2003 to the 1000 cases in 2008–09, were initiated by a traveller who was not detected by the health system for over four weeks. By the time Queensland Health knew about the ignition point for the outbreak, several other people had been infected, and had spread the virus throughout the Cairns area.

Dengue control involves careful synchronisation of a multi-disciplinary team of public health nurses, epidemiologists, entomologists and mosquito control experts, and health promotion workers. Loss of any part of this team can seriously impact the overall success of a control program.

While some control methods in development hold some promise – the *Wolbachia* bacteria that blocks dengue infection in mosquitoes, for instance, and Sumitomo's smokeless mosquito coil that repels and kills mosquitoes in the house – they're still some years away from becoming available.

Given the increase in dengue activity overseas, the need to support current dengue infrastructure has never been greater.

What is herpes?

Dyani Lewis, *Sexual health researcher, University of Melbourne*

When it comes to sexual health, the virus that causes those tingling blisters and angry sores of genital herpes is often the most reviled and feared.

Most cases of genital herpes in humans are caused by the herpes simplex virus 2 (HSV-2), which affects around 12 per cent of Australian adults.

With a reputation for causing unpredictable and unsightly outbreaks in the nether regions and the fact that it stays with you for life, it's little wonder people are often reluctant to divulge that they have the condition. Telling prospective partners can be more excruciating than the condition itself.

Whereas HSV-2 is responsible for most cases of genital herpes, its close cousin, HSV-1, tends to stay above the belt, causing facial cold sores.

HSV-1 and HSV-2 belong to a large family of viruses that we and other animals have been evolving alongside for millennia – the herpesviruses. This nasty family of ultramicroscopic pathogens has evolved to cause brain-swelling encephalitis in cattle, respiratory disease and paralysis in horses and conjunctivitis in goats. Birds get herpes, cats get herpes. Even kangaroos get herpes.

In the late 1960s, researchers observed that HSV-1 and HSV-2 viruses clearly occupied their own distinct territories on the human body. This is still largely true, although the two viruses are increasingly squatting on each other's turf. In 1994, less than 30 per cent of Australian genital herpes cases were caused by the (usually) facial HSV-1, but by 2006, the number was over 40 per cent.

The two viruses are unusual in having evolved to occupy distinct ecological niches on the same host. In our primate relatives, and presumably our distant ancestors, a single virus infects both mouth and genitals. This comes from a lifestyle of frequent genital inspection and oral sex between males and females, and a compact body that allows for self-grooming and auto-fellatio. There was never much distance between face and privates for our forebears.

But gradually, through changes in behaviour and habit, our genitals and mouths became isolated, allowing HSV-1 and HSV-2 to become the genetically distinct viruses that they are today. Walking upright deprived us of the kind of flexibility we more often see in dogs and cats that indulge in self-grooming. And our ancestors' proclivity for oral sex was replaced with a preference for face-to-face sex and kissing, keeping mouths with mouths and genitals with genitals.

Today, with sexual norms once again changing, our predilection for oral sex is more than likely behind the increasing rates of HSV-1 below the belt, and HSV-2 above.

As many a hapless victim will attest, the first outbreak of genital herpes is usually the worst. The virus infects the delicate skin and mucous membranes of the genitals, causing watery blisters four to seven days after exposure. Blisters can occur on the penis in men, and labia, clitoris and vulva in women, but infections can also occur in the anus or on the buttocks and inner thighs.

In addition to the blisters, which eventually harden and heal over a period of two to three weeks, the initial infection can cause fever, headache, muscle pain, swollen lymph nodes and fatigue.

All herpesviruses establish lifelong infections in their host, lurking out of sight of the immune system during periods of latency and bursting forth in orgies of viral reactivation and replication that cause fresh bouts of the tell-tale blisters.

The alpha subfamily of herpesviruses, of which HSV-1 and HSV-2 are members, are neurotropic viruses – they pitch camp in nerve cells during latency. After the initial infection, virus particles travel away from the sensory nerve endings at the skin surface, along the spindly nerve axon to the nucleus in the bulbous nerve body where they lie dormant. During latency, the virus is not only able to evade immune detection, but it also prevents the nerve cell from dying, ensuring that its residence will be a long one.

Reactivation is infuriatingly unpredictable – though factors such as stress and illness can provoke flare-ups. Fresh outbreaks can occur multiple times per year for some, and hardly ever for others. Fortunately, the frequency of outbreaks usually decreases over time.

While viral shedding – the release of viral particles capable of infecting a partner – is greatest during active outbreaks, it is now recognised that transmission from one person to another can occur at any time. The latency period is more like a leaky tap than a closed faucet when it comes to shedding.

Combine this with the fact that shedding can occur in areas not covered by condoms, and that researchers have so far failed to develop an effective vaccine, and it's easy to see why herpes remains a frustratingly difficult infection to control.

Fortunately, up to 80 per cent of people who contract HSV-2 – or HSV-1, for that matter – remain completely asymptomatic. For the remaining 20 per cent, antiviral drugs can lessen the duration and severity of outbreaks, and although embarrassing, the virus does not cause long-term damage. We have lived with HSV-1 and HSV-2 for at least the last 8 million years, and it seems likely we will coexist for a while yet.

What is stuttering?

Nan Bernstein Ratner, *Professor and Chair, Department of Hearing and Speech Sciences, University of Maryland at College Park*

For the one in a hundred adults worldwide who stutter, everyday tasks like picking up a phone, asking for directions, or ordering food in a restaurant can be incredibly difficult.

Stuttering is even more common in young children: as many as 4 per cent of children go through a phase where they repeat or prolong sounds or words, or get 'stuck' trying to talk. Stuttering typically emerges between the ages of two and four, after children have already been speaking normally. As with many other childhood conditions, 80 per cent of stuttering goes away, typically within two years after it first appears.

At this point, we don't know if very young children's recovery from stuttering is aided by therapy, as therapies for stuttering in preschoolers don't achieve a significantly higher success rate than the reported rate of spontaneous, untreated recovery.

Why does a person stutter?

Nobody knows what causes stuttering, but some hypotheses are increasingly being disproved while others gain support.

The common misconception that stressful events or unresolved psychological problems in young childhood cause stuttering, for instance, has literally no evidence base. This was a popular theory in the first half of the 20th century and was explored in the film *The King's Speech* to explain why King George VI started to stutter. But stuttering is not improved by psychological therapies, which indicates it does not have a psychological cause.

At the same time, genetics research, sophisticated brain imaging and motor coordination research support the likelihood that stuttering is caused by trouble integrating the brain 'circuits' that control language formulation and translating of spoken messages into smooth motor actions. This suggests that a person is genetically predisposed to stutter.

Dennis Drayna, a geneticist at the American National Institutes of Health, has identified several plausible candidates for a gene or multiple genes, or genetic mutations, that appear to disproportionately affect people who stutter (Drayna and Kang 2011; Kang and Drayna 2012).

In a series of studies, Luc De Nil and colleagues at the University of Toronto have demonstrated that people who stutter take more time to learn novel motor tasks, make more errors on such tasks and have performance profiles that suffer when they are asked to complete two tasks at the same time (Bauerly and De Nil 2011; Smits-Bandstra and De Nil 2009).

Researchers Anne Smith and Christine Weber-Fox at Purdue University have been able to use the same group of children and adults who stutter to show that they tend to demonstrate less stable motor coordination while learning a new activity, such as tapping a rhythm. They have also shown that speech motor coordination is disproportionately affected by tasks requiring more sophisticated language skills (Smith, Goffman, Sakisekaran, Weber-Fox 2012; Sakisekaran, Smith, Sadagopan, Weber-Fox 2010; Smith, Sadagopan, Walsh, Weber-Fox 2010).

Weber-Fox's lab study also demonstrated that people who stutter display very subtle differences in how the brain processes language, even when listening to speech input, as opposed to talking (Hampton and Weber-Fox 2008; Weber-Fox and Hampton 2008). Such findings point to a very complex communication disorder which combines genetic predispositions with difficulties in integrating across many learning, motor and language systems, which may explain why it has not been easy to find a simple, single explanation for stuttering.

Stuttering is a very handicapping condition that has an impact on social interactions, vocational aspirations and even educational achievement. That's why it's important to seek out good therapy, even for very young children, if they are discomforted by their speaking difficulties.

Many two-year-olds who have trouble speaking fluently seem oblivious of their problem and only their parents are concerned. Other toddlers, however, openly express that they are 'stuck', show signs of physical frustration, or start avoiding words that have caused them difficulty in the past. Any of these negative reactions by the child are reasons to seek immediate help, to make speaking easier and less frustrating to the child.

Therapies that can help

There are several documented options for improving fluency and for navigating moments when speech doesn't come easily.

For older children and adults, the two main approaches teach ways to help people speak more fluently, such as gentler use of voice and the articulators: the tongue and lips. Alongside this, speech therapists teach ways in which the person who stutters can 'slide' more easily out of a moment of stuttering with less obvious struggle and blockage.

For both types of therapy, it may be useful to add a series of components that deal with the speaker's fears about speaking and stuttering – which can be counterproductive to using any new skills learnt in therapy – and 'unlearning' any maladaptive strategies.

Maladaptive strategies, such as trying to force or push 'stuck' words out or gulping air before speaking, have sometimes been taught by well-meaning parents or friends. Stutterers are often advised by those closest to them to 'take a deep breath and try again' when they see them having trouble. This is particularly common advice for children. But we now know it is likely to be counterproductive.

For very young children, a program called Lidcombe, developed at the Australian Stuttering Research Centre in Sydney, has shown to be an effective way to more quickly move preschoolers who stutter into fluency. This is achieved by partnering speech-language pathologists with the child's parents to create a home-based plan of intervention.

While no single therapy has shown to be the 'best', these techniques can help people of any age to speak more fluently, with less struggle and frustration.

References

Bauerly KR, De Nil LF (2011) Speech sequence skill learning in adults who stutter. *Journal of Fluency Disorders* **36**(4), 349–360.

Drayna D, Kang C (2011) Genetic approaches to understanding the causes of stuttering. *Journal of Neurodevelopmental Disorders* 3(4), 374–380.

Hampton A, Weber-Fox C (2008) Non-linguistic auditory processing in stuttering: evidence from behavior and event-related brain potentials. *Journal of Fluency Disorders* 33(4), 253–273.

Kang C, Drayna D (2012) A role for inherited metabolic deficits in persistent developmental stuttering. *Molecular Genetics and Metabolism* **107**(3), 276–280.

Sasisekaran J, Smith A, Sadagopan N, Weber-Fox C (2010) Nonword repetition in children and adults: effects on movement coordination. *Developmental Science* 13(3), 521–532.

Smith A, Goffman L, Sasisekaran J, Weber-Fox C (2012) Language and motor abilities of preschool children who stutter: evidence from behavioral and kinematic indices of nonword repetition performance. *Journal of Fluency Disorders* 37(4), 344–358.

Smith A, Sadagopan N, Walsh B, Weber-Fox C (2010) Increasing phonological complexity reveals heightened instability in inter-articulatory coordination in adults who stutter. *Journal of Fluency Disorders* **35**(1), 1–18.

Smits-Bandstra S, De Nil L (2009) Speech skill learning of persons who stutter and fluent speakers under single and dual task conditions. *Clinical Linguistics & Phonetics* 23(1), 38–57.

Weber-Fox C, Hampton A (2008) Stuttering and natural speech processing of semantic and syntactic constraints on verbs. *Journal of Speech, Language, and Hearing Research* **51**(5), 1058–1071.

Climate and energy

'Men may change their climate, but they cannot change their nature. A man that goes out a fool cannot ride or sail himself into common sense.'

Joseph Addison

'Today the network of relationships linking the human race to itself and to the rest of the biosphere is so complex that all aspects affect all others to an extraordinary degree. Someone should be studying the whole system, however crudely that has to be done, because no gluing together of partial studies of a complex nonlinear system can give a good idea of the behavior of the whole.'

Murray Gell-Mann

'Limits of survival are set by climate, those long drifts of change which a generation may fail to notice. And it is the extremes of climate which set the pattern. Lonely, finite humans may observe climatic provinces, fluctuations of annual weather and, occasionally may observe such things as "This is a colder year than I've ever known." Such things are sensible. But humans are seldom alerted to the shifting average through a great span of years. And it is precisely in this alerting that humans learn how to survive on any planet. They must learn climate.'

Frank Herbert, *Children of Dune*

'Energy is liberated matter, matter is energy waiting to happen.'

Bill Bryson, *A Short History of Nearly Everything*

Climate modes and drought

James Risbey, *Researcher, Marine and Atmospheric Research, CSIRO*

While most people now understand that the enhanced greenhouse effect means a much warmer climate, understanding and communicating the significance and context of weather and climate extremes remains a challenge.

Greenhouse climate change is forced by the build-up of greenhouse gas concentrations in the atmosphere. The climate system responds to this forcing by inexorable increases in temperature and by changing rainfall and circulation patterns. That response is superimposed on the natural fluctuations of climate, and is expressed through the natural modes of the system.

By 'mode' here we mean a preferred circulation pattern. A crude metaphor for a related set of modes might be a bicycle with several gears. Each gear is a preferred mode, whereas the spaces between gears are not preferred. The bicycle is mostly in one of the gears/modes, with occasional transitions between the gears. The climate system has a range of different 'bicycles' describing different types of modes that operate on different time and space scales.

Greenhouse climate change does not present a brand new way for the climate system to operate. The system still has the same ways of expressing variability through atmosphere and ocean circulation modes. The system in which those modes occur is warming. That does not make the modes go away, but may change their behaviour, perhaps by making some modes more active and others less so.

In the metaphor of our bicycle, if we ride out of the flats and into the hills, we still have all the same gears, but we would use the gears differently in the hills. We might prefer to spend more time in the lower gears, for example.

Some commentators argue that recent floods imply we are not undergoing greenhouse climate change, just inter-annual and decadal variability as manifested by La Niña and the Pacific Decadal Oscillation respectively.

They point to natural modes of rainfall variability, as if the existence of these modes ruled out any role for greenhouse climate change. La Niña and decadal variability are intrinsic modes and expressions of the climate system, but that fact does not constitute evidence against greenhouse climate change.

We continue to have La Niña and El Niño events in a changing climate. Indeed, greenhouse climate change would express itself through the natural modes of the system such as La Niña and El Niño in the tropics and blocking (the tendency for persistent high pressure systems to form) in mid-latitudes. The climate system might change the frequency, length, position or intensity of these natural modes in a warmer climate, but the modes themselves will persist.

Similarly, the climate system generates multi-decadal variation in circulation patterns because high frequency weather can generate a low frequency response in the ocean-atmosphere system. To point to such variability is not evidence against greenhouse climate change. A warmer climate system still has multi-decadal variability.

Such variability simply makes it a bit harder to extract and identify the climate change signal. Wet runs of years and dry runs of years will both continue, though the baseline about which these occur may change. If the rainfall baseline drops to lower averages (as seems plausible for southern Australia, based on the research), the dry runs will have larger impacts on the environment.

Multi-decadal variability is a feature of all aspects of the climate system and is not just a property of rainfall. Regional and global temperatures also undergo multi-decadal fluctuations, which is why we don't expect each year, or even decade, to set a new record for warming.

Some climate change critics have tried to have natural variability both ways to suit themselves. They argue that rainfall changes must be assessed over multiple decades on the one hand, but claim inconsistently (and incorrectly) that any brief interruption in the upward march of temperature is evidence against greenhouse climate change. Global surface temperature is always going to bounce around on top of the warming trend.

We do have high confidence that temperatures will continue to rise under greenhouse forcing. Some critics imply that increasing temperatures have no impact on the severity of drought and dispute the impact of those rising temperatures on the water balance during dry runs.

They make this argument by noting that temperatures are higher during droughts than wet periods due to the drying of soils, which changes the way energy is partitioned between evaporation and heating. This is true, is not in dispute, and has no bearing on the greenhouse role.

The temperature changes we are concerned about are not the natural increases that always occur during drought periods. Rather, we are concerned about the inexorable greenhouse-induced rise in global and regional temperature that raises temperature everywhere. Droughts will continue to raise local temperatures, but that rise will occur on top of a warmer baseline. The relevant question is whether droughts are exacerbated when they occur in a warmer climate due to the greenhouse-induced rise in temperature baseline.

If temperature were the only thing to change, then the atmospheric demand for water (termed the potential or Priestley-Taylor evaporation) would increase during droughts in warmed climates. That result is a consequence of the dependence of potential evaporation on temperature. Increases in potential evaporation increase aridity and reduce the amount of available water during droughts. Changes to other variables may oppose or support these changes, but the role of temperature increases will be to increase potential evaporation.

Rainfall is set to undergo a variety of different changes across Australia. While the net impact of greenhouse-induced rainfall changes on drought depends on the precise regional changes in rainfall and on a better quantification of water loss processes, it is very likely that the contribution from greenhouse temperature increases will be to make each drought more stressful than it would otherwise have been.

Greenhouse climate change will generally be expressed through changes in the statistics of the preferred climate modes, but not by the extinction of those modes as such. The occurrence of a given mode or event such as a run of wet years doesn't constitute evidence against greenhouse climate change. It is just an expression of a mode that is always present.

On the other hand, the spatial and temporal fingerprint of the century-long changes in temperature is evidence of greenhouse-forced climate change, and that in turn is changing the system in which the modes operate. We can expect changes in the statistics of wet runs (floods) and dry runs (droughts), superimposed on a warming baseline that tends to exacerbate both extremes of the hydrological cycle.

Storing renewable energy

Maria Skyllas-Kazacos, *Professor Emeritus, School of Chemical Engineering, University of NSW*

Storage is one of the highest technological barriers to the spread of renewable energy. When the sun is shining, the tide turning or the wind howling, how do we collect that energy and keep it to use when generation is down?

There are many different types of energy storage technology available or under development. But each technology has some inherent limitations or disadvantages that make it practical or economical for only a limited range of applications.

Some technologies, such as flywheels, pumped hydro and compressed air, are mechanical. These have low energy efficiencies and slow response times. Pumped hydro and compressed air storage systems are also restricted by special geological and geographical requirements, high investment costs and long construction times.

Electrochemical energy storage systems – batteries – offer many benefits and advantages compared with other forms of energy storage.

Amongst the different types of battery technologies currently available, the ones receiving the most attention for large-scale energy storage applications are:

- lead-acid;
- lithium-ion;
- sodium-sulphur; and
- flow batteries.

Lead-acid batteries

Lead-acid batteries are low in cost. But their application for large-scale energy storage is limited by their short cycle life and limited rechargability.

CSIRO recently developed the UltraBattery. This hybrid energy storage device integrates a supercapacitor (a passive two-terminal electrical component that stores a large amount of energy that can be released very quickly) and a lead-acid battery. The UltraBattery can be charged much faster than conventional lead-acid batteries.

This type of battery is well suited to hybrid vehicles and could be used to smooth out short-term power fluctuations in wind turbines (though this ability has not yet been demonstrated).

Lithium-ion (Li-ion) batteries

The main advantages of Li-ion batteries, compared with other advanced batteries, are:

- high energy density – they can store a large amount of energy in a smaller physical space;
- high efficiency; and
- relatively long cycle life, but still not adequate for many applications.

Lithium-ion batteries perform well and are widespread in portable devices, such as mobile phones.

But several challenges need to be overcome if they're to be used in large-scale grid-connected applications.

The main hurdle is their high cost (more than $600/kWh). Li-ion batteries need special packaging and internal overcharge protection circuits to overcome safety issues that can lead to fires and potential explosions. Some of these safety issues have been addressed with the use of new electrode materials that operate at much lower voltages, but this reduces their energy density (Li-ion's main advantage).

Several companies are also working to reduce the manufacturing cost of Li-ion batteries. The electric vehicle industry is pushing this development.

Sodium-sulphur (Na/S) batteries

The sodium-sulphur (Na/S) battery has a liquid (molten) sulphur positive electrode and liquid sodium negative electrode separated by a solid beta alumina ceramic electrolyte. Na/S battery cells are about 89 per cent energy efficient.

The battery has to be kept at above 300°C to prevent electrolyte freezing and irreversible damage to the cells. These batteries also have safety issues.

The main difficulty with Na/S technology is producing the solid beta alumina tubes that act as both separator and solid electrolyte. These are difficult to mass produce at an affordable cost.

Despite its high cost, Na/S technology has been extensively implemented in a large number of energy storage field trials and demonstrations around the world.

Flow batteries

Flow batteries are a cross between a conventional battery and a fuel cell. They have up to 80 per cent energy efficiency. In contrast to conventional batteries, the power generation (kW) and energy storage capacity (kWh) can be independently varied to suit the job the battery is doing.

Flow batteries are the cheapest energy storage technology available for applications requiring storage of more than four hours, such as large-scale renewable energy storage.

Of the different types of flow battery chemistries that have been explored, only two types – zinc bromine (Zn/Br) and all-vanadium redox (VRB) – have reached commercial fruition.

Three companies are developing the Zn/Br battery commercially. RedFlow, an Australian company, has been working on an improved battery. This is a 120 kVA, 240 kWH grid-connected energy storage system designed to store off-peak electricity for time-shifting and network stabilisation.

VRB was pioneered at the University of New South Wales. The largest VRB installation to date is a 4 MW/6 MWh demonstration system built by Sumitomo Electric Industries to store energy at a windfarm on the island of Hokkaido in Japan. Up to 200 000 charge-discharge cycles were demonstrated over the

A 200 kW/400 kWh VRB system. Courtesy Paul McDermott and Gildemeister

three-year life of the project, as well as very high energy efficiencies and fast response times that enabled output power stabilisation for the wind turbines.

The cost per kWh of generated energy of a VRB system can be less than half that of an equivalent lead-acid battery system for storage capacities in excess of four hours. This makes the VRB one of the cheapest energy storage technologies for large-scale renewable energy storage.

Several companies are now commercially manufacturing VRB systems, and VRB is likely to be one of the leading battery technologies in the expanding global energy storage market being fuelled by the push towards renewables and the smart grid.

What are biofuels?

Daniel Tan, Senior Lecturer, Agriculture, University of Sydney

Since the beginning of civilisation, humans have depended on organic materials – or 'biomass' – for cooking and heat. Many developing countries in Asia and Africa still do. Biofuel or bioenergy is the chemical energy contained in biomass that can be converted into direct, useful energy sources using biological (including food digestion), mechanical or thermochemical processes.

Current liquid biofuels (bioethanol and biodiesel) are mainly produced from first generation feedstocks (such as sugarcane, maize, rapeseed) and constitute only a small fraction (1 per cent) of present transportation energy. Second generation biofuels will come from dedicated perennial energy crops, such as miscanthus, switchgrass, agave, and pongamia. In the near future, hydrogen gas may be produced from algae, bacteria or artificial photosynthesis to fuel hydrogen-cell powered cars.

Liquid biofuels

Liquid biofuels are most familiar to us. Bioethanol is a substitute for gasoline and biodiesel is a substitute for diesel.

At present most cars have internal combustion engines which can only burn liquid fuels. Other types of engines, such as electric and hydrogen fuel cells, are being developed. These produce power in much the same way as a battery, but run continuously as long as the fuel supply is maintained. In the meantime, though, liquid biofuels are the transition renewable alternative to fossil fuels for transport.

According to the International Energy Agency, liquid biofuels account for only 2 per cent of total bioenergy, and they are mainly significant in the transportation sector. Transportation accounts for 28 per cent of global energy consumption and 60 per cent of global oil production. Liquid biofuels supplied only 1 per cent of total transport fuel consumption in 2009.

Globally, liquid biofuels can be classified into three main production sources: maize ethanol from the United States, sugarcane ethanol from Brazil and rapeseed biodiesel from the European Union. There is also a small quantity of palm oil biodiesel from Indonesia.

In 2010, Brazil and the United States produced 90 per cent of the 86 billion litres of global bioethanol and the European Union produced 53 per cent of the 19 billion litres of biodiesel.

In Australia, biodiesel is being produced from used cooking oil (an agricultural by-product), tallow and canola seed; and bioethanol is produced from sugarcane molasses, grain sorghum and waste wheat starch.

Not so sustainable

The first generation of biofuels produced from starches, sugars and oils of agricultural food crops, including maize, sugarcane, rapeseed (including canola) and soybean have faced disfavour for competing with food and feed production.

Hypothetically, if all the main cereal and sugar crops (wheat, rice, maize, sorghum, sugarcane, cassava and sugar beet), representing 42 per cent of global cropland, were to be converted to ethanol, this would correspond to only 57 per cent of total petrol use in 2003, and leave no cereals or sugar for human consumption (although the reduced sugar in the human diet would have health benefits).

These first generation biofuels also have large carbon and water footprints. Greenhouse gas emissions during agricultural production of biofuel crops contribute 34–44 per cent of the greenhouse gas balance of maize ethanol in the United States and more than 80 per cent of that of pure vegetable oils. In general the water footprint of biofuels is two to five times greater than the water footprint of fossil fuels.

Clearing undisturbed native ecosystems such as rainforest, savanna and grassland for biofuel production also increases net greenhouse gas production because of the change in land use.

The way forward

Because of food and energy security concerns, many countries are promoting biofuel crops that can be grown on land not suited for food production, so the two systems are complementary rather than competitive.

The term 'second generation biofuels' refers to the range of feedstocks (dedicated energy crops such as miscanthus, switch-grass, pongamia, agave, Indian mustard, sweet sorghum, algae and carbon waste). It also includes conversion technologies such as:

- fast pyrolysis (heating of biomass in the absence of oxygen);
- supercritical water (water kept above 374°C, at which point it loses its hydrogen bonds); and
- gasification (heating organic materials at high temperature, without combustion, to produce a synthetic gas).

Thermo-chemical refining technologies such as Fischer-Tropsch methods (processes that convert a mixture of carbon monoxide and hydrogen into liquid hydrocarbons) that can be used to convert biomass into useful fuels are also covered under this umbrella term.

For example, Swiss company Clariant opened a plant in Germany in 2012 that can produce up to 1000 tons of cellulosic ethanol from 4500 tons of wheat straw.

Hydrogen is considered a third generation biofuel, when it is produced from biomass by algae or enzymes. It contains three times the energy of petrol on a mass basis and its combustion produces only water. Future technological breakthroughs are needed before hydrogen can be produced economically.

One method of hydrogen production is artificial photosynthesis. This involves mimicking natural systems using molecular photocatalytic systems (in which light reactions are speeded up by the use of a catalyst) to enable water oxidation and hydrogen production.

Artificial photosynthesis was only an academic pursuit until the development of the first practical artificial leaf. It is claimed to be potentially 10 times more efficient in photosynthesis than a natural leaf; however, commercialisation of artificial photosynthesis is yet to be proven.

What is photovoltaic solar energy?

Andrew Blakers, Director, Centre for Sustainable Energy Systems (CSES), ANU

There are two main types of solar energy technology: photovoltaics (PV) and solar thermal. Solar PV is the rooftop solar you see on homes and businesses – it produces electricity from solar energy directly. Solar thermal technologies use the sun's energy to generate heat, and electricity is generated from that.

Australia receives thousands of times more energy from the sun each year than from all fossil fuel use combined. At the moment we directly use only about one millionth of this as commercial energy.

How it works

Solar photovoltaic is an elegant technology which produces electricity from sunlight without moving parts.

In a photovoltaic cell, sunlight detaches electrons from their host silicon atoms. Tiny packets of light energy called photons are captured by electrons, and impart enough energy to kick the electron free of its host atom. Near the upper surface of the cell is a 'one-way membrane' called a pn-junction. The pn-junction is formed by diffusing tiny quantities of phosphorus to a depth of about one micrometre into a thin wafer of silicon.

When a free electron crosses the pn-junction it cannot easily return, causing a negative voltage to appear on the surface facing the sun (and a positive voltage on the rear surface). The front and rear surfaces can be connected together via an external circuit in order to extract current, voltage and power from the solar cell.

Solar cells are packaged behind glass to form photovoltaic modules, which have typical service lives of 20 to 40 years.

In many circumstances, photovoltaic modules mounted on building roofs can produce as much electricity as the building consumes. A typical module will generate about 200 kW hours of alternating current per square metre per year, so a collector area of 25–50 m^2 is needed to power a reasonably energy-efficient Australian house. Such a house exports more electricity to the grid during the day than it imports at night.

An additional 10 m^2 is required to offset the annual greenhouse gas emissions of a fuel-efficient car emitting 0.2 kg of carbon dioxide per km and driving 10 000 km per year.

A changing market

The world solar market is dominated by photovoltaics, and most of the world's PV market is serviced by crystalline silicon solar cells. Until now PV has found widespread use in niche markets such as consumer electronics, remote area power supplies and satellites.

In recent years there have been dramatic falls in the cost of solar PV and the industry has expanded immensely. Panel prices are now below $1000 per kilowatt and system prices are $2000–$3000 per kilowatt. Solar PV electricity is now less expensive than both domestic and commercial retail electricity from the grid. It is approaching cost-competitiveness with wholesale conventional electricity in many places.

The cost of photovoltaic systems can be confidently expected to continue to decline for decades. Current worldwide PV module sales are 30–40 gigawatts per year (approximately equal to the power capacity of the Australian electricity system). In 2012 about one gigawatt was installed in Australia: nearly one million Australian houses have photovoltaic systems on their roofs. A long-term growth trend is likely to continue because the cost of solar electricity is well below the retail commercial and domestic tariff everywhere in Australia.

Problems in the industry

In recent years photovoltaic production capacity has surged far ahead of demand, which has led to an intense price war and a wave of bankruptcies of high profile PV companies. Technological innovation is being squeezed by the need of companies to focus on survival. A profound industry shakeout and consolidation is under way. The winners will emerge into a trillion dollar per year industry.

The future of the solar PV industry is driven by the fundamental equation that PV is little constrained by environmental considerations, material supply, land requirements, security considerations, or indeed anything other than price – and price is now (or will soon be) competitive nearly everywhere.

The solar energy revolution

Grid parity at retail level for photovoltaics has already been achieved for most of the world's businesses and population. In Australian cities, rooftop solar systems typically produce electricity at a cost of half to two-thirds of what domestic and commercial retail electricity usually sells for. This is leading to rapid growth in sales in the residential and commercial sectors without the need for subsidies.

Some people are concerned that more use of variable renewables like solar and wind power will make it hard for the grid to cope – they provide power intermittently but our demand is more constant. However, this concern is misplaced; for example, South Australia now procures nearly 30 per cent of its electricity from wind and rooftop solar and has avoided major problems.

This is because many options are available for managing fluctuations in the amount of energy provided by wind and photovoltaics, including shifting demand from night to day (the opposite of what is done at present), using a range of renewable energy technologies and mass storage including pumped hydroelectric storage – the method used by Snowy Hydro among others. This is by far the leading energy storage technology worldwide.

What we know and don't know about climate change

Kevin Hennessy, Principal Research Scientist, Marine and Atmospheric Research, CSIRO

Most Australians believe that climate change is real and want to learn more about it, but the debate in the media and on the internet makes it difficult for lay people to know who and what to believe.

There are uncertainties in climate science, as in any scientific field, and scientists are quite open about these. But these uncertainties are often misrepresented.

In the claims and counterclaims by various climate change experts and other commentators, one yardstick that can be used is whether the 'science' being put forward has passed the scrutiny of peer-review before being published.

Peer review is a process in which relevant experts assess the competence, significance and originality of the research.

There are several assessments of peer-reviewed climate change science:

- The Intergovernmental Panel on Climate Change (2007): 'Climate Change 2007 Synthesis Report'. Download available at http://www.ipcc.ch/pdf/assessment-report/ar4/syr/ar4_syr_spm.pdf.
- The Royal Society (2010): 'Climate Change: A Summary of the Science'. Download available at http://royalsociety.org/policy/publications/2010/climate-change-summary-science/.

- The Australian Academy of Science (2010): 'The Science of Climate Change: Questions and Answers'. Download available at http://www.science.org.au/reports/climatechange2010.pdf.
- The Climate Change Commission (2011): 'The Critical Decade'. Download available at http://climatecommission.gov.au/report/the-critical-decade/.
- World Meteorological Organization (2013): 'A Summary of Current Climate Change Findings and Figures'. Download available at http://www.wmo.int/pages/mediacentre/factsheet/documents/Climate-Change-Info-Sheet-136_fr.pdf.

Within each of these reports there is a distinction between science that is robust and science that is relatively uncertain.

Chapter 6 of the IPCC (2007) Synthesis Report lists 21 robust findings and 18 key uncertainties. These are grouped and summarised below.

Robust findings: the things we know
- There is clear evidence for global warming and sea level rise.
- Changes that are being observed in many physical and biological systems are consistent with warming.
- Due to the uptake of anthropogenic CO_2 since 1750, ocean acidity has increased.
- Most of the global average warming over the past 50 years is very likely due to anthropogenic greenhouse gas increases.
- Global greenhouse gas emissions will continue to grow over the next few decades, leading to further climate change.
- Due to the time scales associated with climate processes and feedbacks, anthropogenic warming and sea level rise would continue for centuries even if greenhouse gas emissions were to be reduced sufficiently for atmospheric concentrations to stabilise.
- Increased frequencies and intensities of some extreme weather events are very likely.
- Systems and sectors at greatest risk are ecosystems, low-lying coasts, water resources in some regions, tropical agriculture, and health in areas with low adaptive capacity.

- The regions at greatest risk are the Arctic, Africa, small islands and Asian and African mega-deltas. Within other regions (even regions with high incomes) some people, areas and activities can be particularly at risk.
- Some adaptation is under way, but more extensive adaptation is required to reduce vulnerability to climate change.
- Unmitigated climate change would, in the long term, be likely to exceed the capacity of natural, managed and human systems to adapt.
- Many impacts can be reduced, delayed or avoided by mitigation (net emission reductions). Mitigation efforts and investments over the next two to three decades will have a large impact on opportunities to achieve lower greenhouse gas stabilisation levels.

Key uncertainties: the things we're not sure about

- Observed climate data coverage remains limited in some regions.
- Analysing and monitoring changes in extreme events is more difficult than for climatic averages because longer datasets with finer spatial and temporal resolutions are required.
- Effects of climate changes on human and some natural systems are difficult to detect due to adaptation and non-climatic influences.
- Difficulties remain in reliably attributing observed temperature changes to natural or human causes at smaller than continental scales.
- Models differ in their estimates of the strength of different feedbacks in the climate system, particularly cloud feedbacks, oceanic heat uptake and carbon cycle feedbacks.
- Confidence in projections is higher for some variables (such as temperature) than for others (such as precipitation), and it is higher for larger spatial scales and longer averaging periods.
- Direct and indirect aerosol impacts on the magnitude of the temperature response, on clouds and on precipitation remain uncertain.

- Future changes in the Greenland and Antarctic ice sheet mass are a major source of uncertainty that could increase sea level rise projections.
- Impact assessment is hampered by uncertainties surrounding regional projections of climate change, particularly precipitation.
- Understanding of low-probability/high-impact events and the cumulative impacts of sequences of smaller events is generally limited.
- Barriers, limits and costs of adaptation are not fully understood.
- Estimates of mitigation costs and potentials depend on uncertain assumptions about future socio-economic growth, technological change and consumption patterns.

Do we know the world is warming due to human activity?

The IPCC statement most often challenged by so-called sceptics is 'Most of the global average warming over the past 50 years is very likely due to anthropogenic greenhouse gas increases.' Those who are keen to dig deeper into the peer-reviewed literature on this issue can read more:

- Chapter 9 of the IPCC (2007) Working Group 1 report.
- Easterling and Wehner (2009) 'Is the climate warming or cooling?'
- Stott *et al.* (2010) 'Detection and attribution of climate change'.
- Kaufmann *et al.* (2011) 'Reconciling anthropogenic climate change with observed temperature 1998–2008'.
- Trenberth (2012) 'Framing the way to relate climate extremes to climate change'.

Some of the other common issues raised about climate change science have been addressed by CSIRO. In partnership with the Australian Bureau of Meteorology, CSIRO has also released several 'State of the Climate' reports. You can download a copy of the 2012 report at http://www.csiro.au/en/Outcomes/Climate/Understanding/State-of-the-Climate-2012.aspx.

Navigating the maze of information about climate change science is challenging for a layperson. Recent assessments of the peer-reviewed literature put this into perspective.

There are many robust findings about the science, and these provide a basis for action through mitigation of greenhouse gases as well as adaptation to reduce our vulnerability to climate change impacts.

While there are also scientific uncertainties that need further research, these don't undermine a compelling scientific case for increased risk management.

References

Easterling DR, Wehner MF (2009) Is the climate warming or cooling? *Geophysical Research Letters* **36**, L08706.

Kaufmann RK, Kauppi H, Mann ML, Stock JH (2011) Reconciling anthropogenic climate change with observed temperature 1998–2008. *Proceedings of the National Academy of Sciences of the United States of America* **108**(29), 11790–11793.

Stott PA, Gillett NP, Hegerl GC, Karoly DJ, Stone DA, Zhang X, Zwiers F (2010) Detection and attribution of climate change: a regional perspective. *WIREs Climate Change* **1**, 192–211.

Trenberth K (2012) Framing the way to relate climate extremes to climate change. *Climatic Change* **12**, 283–290.

Ever wondered?

'Aristotle maintained that women have fewer teeth than men; although he was twice married, it never occurred to him to verify this statement by examining his wives' mouths.'

Bertrand Russell, *The Impact of Science on Society*

'I am not more gifted than the average human being. If you know anything about history, you would know that is so – what hard times I had in studying and the fact that I do not have a memory like some other people do … I am just more curious than the average person and I will not give up on a problem until I have found the proper solution. This is one of my greatest satisfactions in life – solving problems – and the harder they are, the more satisfaction do I get out of them. Maybe you could consider me a bit more patient in continuing with my problem than is the average human being. Now, if you understand what I have just told you, you see that it is not a matter of being more gifted but a matter of being more curious and maybe more patient until you solve a problem.'

Albert Einstein

"'Mercy!" cried Gandalf. "If the giving of knowledge is to be the cure of your inquisitiveness, I shall spend all the rest of my days in answering you. What more should you like to know?"

"The names of all the stars, and of all living things, and the whole history of Middle-Earth and Over-heaven and of the Sundering Seas," laughed Pippin. "Of course! What less?"'

J.R.R. Tolkien, *The Two Towers*

'I set out to discover the why of it, and to transform my pleasure into knowledge.'

Charles Baudelaire

Does luck exist?

Neil Levy, Head of Neuroethics, Florey Institute of Neuroscience and Mental Health, University of Melbourne

Some people seem born lucky. Everything they touch turns to gold. Others are dogged by misfortune.

It's not just people who might be lucky or unlucky – it can be single acts. When the ball hits a post in soccer, the commentators often say the striker was unlucky. We sometimes argue whether an act was lucky or not. I might say your pool shot was lucky. 'Not luck; skill,' you might reply.

Is any of this talk sensible? Is there really such a thing as luck? Do some people have more of it than others (just as some people are better at pool than others)? I think there is a perfectly reasonable way of making sense of talk about luck. But there is no such thing as luck. It isn't a property, like mass, or an object. Rather, to talk about luck is to talk about how things might easily have gone.

This view entails that no-one has luck. We can't truly say of someone they're lucky, meaning they are the kind of person to whom lucky things can be expected to happen.

It has sometimes been suggested that luck exists only if a certain interpretation of quantum mechanics is true: if causality is not 'deterministic'. If physical determinism is true, and there are conditions under which nothing other than what *did* happen *could* have happened, then every event that occurs is entirely predictable (in principle), by someone who knows enough about the universe and its laws.

If indeterministic physics is true, then such predictability is not possible: no one, no matter how much they know, can predict every event that happens, even in principle.

I don't know which interpretation of quantum mechanics is true, but it seems unlikely to me that we need to settle that debate to decide whether some things are lucky. It seems obvious to me that the person who was hit by lightning (on a clear day, if you like) was unlucky, and the person who wins the lotto is lucky.

Here's how I understand luck. I think something is lucky (or unlucky) for a person if two things are true of it: it matters (somehow) to them, and it might easily not have happened. The second condition needs some explanation.

To say that something might easily not have happened is to say that, given how things were at the time just before, the event might well not have occurred. We might think of this in terms of replaying the event. If I set up the pool table again and ask you to retake the shot, we can discover whether your shot was luck or skill. We will need to do it a few times: you might get lucky twice, but you're very unlikely to be lucky 10 times in a row.

If every time you try (roughly) the same shot, you sink it, I will have to concede: that's skill, not luck. But if you can't do it again, you were lucky the first time. Similarly, someone was unlucky to be hit by lightning if it is true that were they to be in similar conditions again, they (probably) would not be hit by lightning. If, on the other hand, lightning is so prevalent around here that any time anyone goes out they get hit, then they weren't unlucky.

If this is right, there can't be lucky or unlucky people. At least, there can't be people who have the property of having lucky events happen to them. Whether I am lucky in doing something depends on how skillful I am at doing things like that. If I'm really good at it, then I am less lucky at succeeding than if I am bad at it.

So, roughly, the more often something happens to someone, the less luck is involved. Of course someone can be lucky or unlucky twice: lightning can strike twice. But the person who is lucky twice, or more, is not a lucky person: their past luck doesn't give us any reason to expect luck in their future.

There is one way in which we can say that someone is lucky or unlucky. Rather than compare an event with what we would expect to happen, given roughly the same circumstances, we might compare a person's circumstances or their traits to what is statistically normal for a group. Using this kind of measure, we can say that someone born severely handicapped is unlucky and someone born into wealth is lucky.

What is the relevant group for this kind of comparison? I don't think there is a single right answer here: it will depend on the context and our aims. For some purposes, a narrower group might be relevant, and for some, a broader. This entails that the same person might be said to be both lucky and unlucky.

Think of the contemporary Australian who loses her job, through no fault of her own. We might say she is unlucky, in comparison with other contemporary Australians. But compared with humanity as a whole, she might be lucky if she remains able to feed and house herself.

This same kind of context sensitivity and relativism is characteristic of luck in events as well. The same event can be lucky and unlucky for a person. Think of someone who misses her flight and takes another one, which then crashes. She is unlucky to be involved in a plane crash, given that she might easily have been on the earlier flight. But if she is the only survivor, she might be lucky, given that everyone else died.

That's why we can find ourselves saying someone who has broken three ribs and both legs is lucky.

Why hypothesis and significance tests ask the wrong questions

Rob Herbert, *Senior Principal Research Fellow, Neuroscience Research Australia*

Empirical science needs data. But all data are subject to random variation, and random variation obscures patterns in data. So statistical methods are used to make inferences about the true patterns or effects that underlie noisy data.

Most scientists use two closely related statistical approaches to make inferences from their data: significance testing and hypothesis testing. Significance testers and hypothesis testers seek to determine if apparently interesting patterns ('effects') in their data are real or illusory. They are concerned with whether the effects they observe could just have emanated from randomness in the data.

The first step in this process is to nominate a 'null hypothesis', which posits that there is no effect. Mathematical procedures are then used to estimate the probability that an effect at least as big as that observed would have arisen if the null hypothesis were true. That probability is called 'p'.

Significance testing

If p is small (conventionally less than 0.05, or 5 per cent) then the significance tester will claim that it is unlikely an effect of the observed magnitude would have arisen by chance alone. Such effects are said to be 'statistically significant'. Sir Ronald Fisher who, in the 1920s, developed contemporary methods for generating p values, interpreted small p values as being indicative of 'real' (not chance) effects. This is the central idea in significance testing.

Hypothesis testing

Significance testing has been under attack since it was first developed. Two brilliant mathematicians, Jerzy Neyman and Egon Pearson, argued that Fisher's interpretation of p was dodgy. They developed an approach called hypothesis testing in which the p value serves only to help the researcher make an optimised choice between the null hypothesis and an alternative hypothesis: If p is greater than or equal to some threshold (such as 0.05) the researcher chooses to believe the null hypothesis. If p is less than the threshold the researcher chooses to believe the alternative hypothesis. In the long run (over many experiments) adoption of the hypothesis testing approach minimises the rate of making incorrect choices.

Critics have pointed out that there is limited value in knowing only that errors have been minimised in the long run – scientists don't just want to know they have been wrong as infrequently as possible, they want to know if they can believe their last experiment!

The most vociferous critic of hypothesis testing was Fisher, who hounded Neyman in print for decades (celebrated mathematical statistician Leonard Jimmie Savage said Fisher 'published insults that only a saint could entirely forgive'). Perhaps largely as a result of Fisher's intransigence, the issues that divided significance testing and hypothesis testing were never resolved.

Statistical inference

Today's scientists typically use a messy concoction of significance testing and hypothesis testing. Neither Fisher nor Neyman would be satisfied with much of current statistical practice.

Scientists have enthusiastically adopted significance testing and hypothesis testing because these methods appear to solve a fundamental problem: how to distinguish 'real' effects from randomness or chance. Unfortunately significance testing and hypothesis testing are of limited scientific value – they often ask the wrong question and almost always give the wrong answer. And they are widely misinterpreted.

Consider a clinical trial designed to investigate the effectiveness of new treatment for some disease. After the trial has been conducted the researchers might ask, 'Is the observed effect of treatment real, or could it have arisen merely by chance?' If the calculated p value is less than 0.05, the researchers might claim the trial has demonstrated the treatment was effective. But even before the trial was conducted we could reasonably have expected the treatment was 'effective' – simply because almost all drugs have some biochemical action and all surgical interventions have some effects on health. Almost all health interventions have some effect, it's just that some treatments have effects that are large enough to be useful and others have effects that are trivial and unimportant.

So what's the point in showing empirically that the null hypothesis is not true? Researchers who conduct clinical trials need to determine if the effect of treatment is big enough to make the intervention worthwhile, not whether the treatment has any effect at all.

A more technical issue is that p tells us the probability of observing the data given that the null hypothesis is true. But most scientists think p tells them the probability the null hypothesis is true given their data. The difference might sound subtle but it's not. It is like the difference between the probability that a prime minister is male and the probability a male is prime minister.

A better approach to statistical inference
There are alternatives to significance testing and hypothesis testing. A simple alternative is 'estimation'. Estimation helps scientists ask the right question, and provides better (more statistically defensible, if not more mathematically rigorous) answers.

Another very different approach is 'Bayesian' analysis. Bayesian statisticians try to quantify uncertainty and use data to modify their certainty about particular beliefs. In many ways Bayesian methods are superior to classic methods but scientists have been slow to adopt Bayesian approaches.

Significance testing and hypothesis testing are so widely misinterpreted that they impede progress in many areas of science. What can be done to hasten their demise? Senior scientists should ensure that a critical exploration of the methods of statistical inference is part of the training of all research students. Consumers of research should not be satisfied with statements that 'X is effective', or 'Y has an effect', especially when support for such claims is based on the evil p.

Why is the sky blue?

Murray Hamilton, *Associate Professor in Physics, University of Adelaide*

A young child looked up in the sky,
And said, 'It's so blue, Mum, but why?'
You see, blue scatters more,
(There's this power of 4),
So it rarely comes straight to your eye.

Author unknown

Most of what is between us and space is air, which is made up of very small molecules. There are also varying amounts of other stuff – aerosols, dust, haze, clouds, smoke and so on.

The light from the sun has to pass through this 'stuff' to get to the surface of the earth. But some of the light does not make it, or if it does arrive at the surface, it gets there indirectly.

Why? Because the light – or more accurately, *part* of the light – is scattered by 'stuff' in the atmosphere.

Visible light is an electromagnetic wave of a rather narrow range of wavelengths (roughly 390–700 nanometres, where 1 nanometre is 1 billionth of a metre) in the complete spectrum. This electromagnetic spectrum spans the very long wavelengths of radio to the extremely short wavelengths of gamma rays.

Within the visible portion of the spectrum red light has a longer wavelength than blue – 650 nm versus 450 nm.

When light is scattered in the atmosphere, the amount of scatter and the angle by which it is scattered depend on the wavelength and the size of the scatterer.

If the scatterer's size is significant compared with the wavelength of the light being scattered, the shape of the scatterer becomes important too.

Molecules are the smallest scatterers, being about a factor of 1000 smaller than the wavelength of visible light. For these molecules – such as nitrogen gas (N_2) which makes up 78 per cent of the atmosphere – the dependence of scattering on wavelength goes as the inverse fourth power of wavelength.

That is, comparing blue with red, we take the wavelength ratio (650/450) and raise that to the fourth power to calculate how much more likely (4.3 times, as it turns out) it is blue that will be scattered than the red.

If you look away from the sun, blue light travelling from the sun through the earth's atmosphere (but not directly towards you) is scattered by the molecules towards your eye.

Thus the sky looks blue because scattering from molecules is much more probable for blue light than red.

There is so much blue light from any particular direction that it completely dominates the light from the stars, which is nonetheless still there.

If you get close enough to space, the sky is black. This is because there is nothing significant up there to scatter the sunlight to your eye.

Conversely, if you look through a clean atmosphere (i.e. with no dust or smoke) towards the sun, a significant amount of the blue light is scattered away from the line of sight, which tends to give the sun a yellowish hue.

This sort of scattering – when the scatterer is considerably smaller than the wavelength of the light – is usually called 'Rayleigh scattering', after 19th century British physicist Lord Rayleigh.

Near sunset, the path sunlight takes on its way through the atmosphere to you is especially long. In this case, so much blue light is lost (i.e. scattered away) that the sun appears orange or even red.

The water droplets and ice crystals that make up clouds are quite large compared with the wavelength of visible light – at least 20 times greater. In this case the scattering of light is strong and nearly independent of wavelength, at least over the visible range.

Because nearly all wavelengths of visible light are scattered, clouds appear white, or varying shades of neutral grey if they are in the shadow of other clouds.

If you consider the wavelength dependence of scattering for medium-sized particles, you find that for a narrow wavelength range the dependence reverses for a bit. That is, in this range – which can be as wide as the visible spectrum – the scattering is *stronger* for longer wavelengths, as opposed to weaker.

Sometimes the size of the particles can be just right for this to happen in the visible spectrum. In this case the sunlight that passes through air with these particles suspended has red light scattered away leaving the sun (or moon) looking bluish.

This is rare, but has been known to happen when volcanoes or bushfires load the atmosphere with particles of just the right size.

In contrast to Earth, the Martian atmosphere is quite dusty, and there the sky tends to be orange, sometimes with blue sunsets. This is because the dust particles are much larger than the carbon dioxide molecules which make up the atmosphere.

The fact that the atmosphere is very thin on Mars means scattering from dust is relatively more important than it would be on Earth.

So next time you're lying in the grass looking up at white clouds float across a stunning blue sky, spare a moment to think about the physics responsible for what you're seeing.

Medical myths

'We look for medicine to be an orderly field of knowledge and procedure. But it is not. It is an imperfect science, an enterprise of constantly changing knowledge, uncertain information, fallible individuals, and at the same time lives on the line. There is science in what we do, yes, but also habit, intuition, and sometimes plain old guessing. The gap between what we know and what we aim for persists. And this gap complicates everything we do.'

Atul Gawande, *Complications: A Surgeon's Notes on an Imperfect Science*

'Medicine is not only a science; it is also an art. It does not consist of compounding pills and plasters; it deals with the very processes of life, which must be understood before they may be guided.'

Paracelsus

'Disease is war with the laws of our being, and all war, as a great general has said, is hell.'

Lewis G. Janes

'The poets did well to conjoin music and medicine, because the office of medicine is but to tune the curious harp of man's body.'

Francis Bacon

Drink eight glasses of water a day

Tim Crowe, Associate Professor Nutrition, Deakin University

We have all heard the popular advice that we should drink at least eight glasses of water a day, so it may be a surprise that this is more myth than fact.

Of course our bodies need water, otherwise we would die from dehydration. But the amount needed is extremely variable and depends on a person's body size, physical activity levels, climate and what types of food they are eating.

Water makes up about 60 per cent of an adult's bodyweight and is an essential nutrient, more important to life than any others.

Water helps regulate body temperature, carries nutrients and waste products throughout the body, is involved in blood transport, and allows many metabolic reactions to occur.

It also acts as a lubricant and cushion around joints, and forms the amniotic sac surrounding a foetus.

It is widely believed that the 'eight glasses' myth was a US Recommended Dietary Allowance dating back to 1945.

The guide said a suitable allowance of water for adults was 2.5 litres a day, but most of this water could be found in prepared foods.

If that last, crucial part is ignored, the statement could be interpreted as clear instructions to drink eight glasses of water a day.

Even a comprehensive search of the scientific literature finds no evidence to support the 'eight glasses a day' advice.

The clear reason that evidence for such prescriptive advice doesn't exist is that a person can get all the water they need without consuming a single glass.

Drinks like soft drink, fruit juice, tea and coffee, milk, and foods like fruit, yoghurt, soups and stews all have appreciable amounts of water that contribute to fluid intake.

Australian dietary recommendations also bust the eight-glass myth. The official Nutrient Reference Values states, 'There is no single level of water intake that would ensure adequate hydration and optimal health for the apparently healthy people in the population.'

Don't be concerned about seeing coffee listed as a fluid – the 'coffee makes you dehydrated' mantra is another myth that needs to be busted.

Drinks such as coffee, tea and cola do have a mild diuretic effect from the caffeine but the water loss caused by this is far less than the amount of fluid consumed in the drink in the first place.

It's only alcoholic drinks which have a dehydrating effect.

So how do you know if you are drinking enough water?

Well. You can check this for yourself every few hours. If your urine is lightly coloured or clear, you're drinking enough. If it's dark, then you should drink more.

How simple is that?

We only use 10 per cent of our brain

Kate Hoy, *Research Fellow/Clinical Neuropsychologist, Monash University*

The thought that most of us only use 10 per cent of our brain is appealing because it means we have a whole lot of untapped potential waiting to be harnessed. Unfortunately, that figure is off by about 90 per cent.

This myth has been variously (mis)attributed to William James, Albert Einstein, and even early neuroscience researchers.

While its exact origins are unclear, popular belief has persisted, and even strengthened, since the 1890s, despite the overwhelming evidence to the contrary.

In his book *Mind Myths: Exploring Popular Assumptions About the Mind and Brain*, neuroscientist Barry Beyerstein discusses seven kinds of evidence that refute the '10 per cent myth'.

The most convincing of these involves the use of brain imaging.

There are numerous brain imaging techniques that allow us to see the activity of the brain. These include Positron Emission Tomography (PET) and functional Magnetic Resonance Imaging (fMRI).

These techniques have revealed that all parts of the brain show some level of activity, except in the case of serious damage.

For example, we recently conducted a PET study which required participants to do nothing, simply rest without ruminating on any one thought. This is known as a resting state study.

Even in this so-called 'resting state' the brain scans revealed widespread areas of metabolic activity – far in excess of 10 per cent.

In an individual PET scan at rest we don't see any black areas (which would indicate inactivity), so the entire brain is showing some level of activity.

It's also possible to see the brain activity which occurs when someone is performing a task.

For example, our group used fMRI to look at the pattern of brain activity occurring when people are engaged in a complex problem – solving the task known as the Tower of London, which tests executive function, particularly planning skills.

We saw increased activity in several areas: activity over and above what is seen in the brain when participants were not engaged in a task.

This type of imaging clearly shows our whole brains are always active, to some degree. When we are engaged in a task, specific areas of the brain will become more active, depending on the demands of that task.

A variation on the brain capacity myth is that we only use 10 per cent of our brain at any one time, depending on the task we're doing.

Yet even the seemingly simple task of tapping your finger on a desk requires brain power far in excess of 10 per cent of your resting state.

Such a task involves coordinated activity from many areas, including the sensory and motor cortices, the occipital and parietal lobes, the basal ganglia, cerebellum and frontal cortex.

So how has the 10 per cent myth managed to persist and even thrive?

One reason may be its popularity in books (*Lest We Remember*), film (*Limitless, The Lawnmower Man*), television (*Heroes, Eureka*) and even self-help literature (*How to Win Friends and Influence People*).

The myth is most often presented in popular culture as a hurdle to overcome: by harnessing the rest of our brain power we will be able to achieve amazing feats of intelligence, creativity and (apparently) telekinetic powers.

So it's not surprising that people continue to believe it's true.

But it's not all bad news. The plasticity of the human brain means it is able to constantly reorganise itself, allowing us to develop new skills and abilities right throughout our life.

Alcohol kills brain cells

Nick Dorsch, Clinical Associate Professor, University of Sydney

Do you ever wake up with a raging hangover and picture the row of brain cells that you suspect have started to decay? Or wonder whether that final glass of wine was too much for those tiny cells, and pushed you over the line?

Well, it's true that alcohol can indeed harm the brain in many ways. But directly killing off brain cells isn't one of them.

The brain is made up of nerve cells (neurons) and glial cells. These cells communicate with each other, sending signals from one part of the brain to the other, telling your body what to do. Brain cells enable us to learn, imagine, experience sensation, feel emotion and control our body's movement.

Alcohol's effects can be seen on our brain even after a few drinks, causing us to feel tipsy. But these symptoms are temporary and reversible. The available evidence suggests alcohol doesn't kill brain cells directly.

There is *some* evidence that moderate drinking is linked to improved mental function. A 2005 Australian study of 7500 people in three age cohorts (early 20s, early 40s and early 60s) found moderate drinkers (up to 14 drinks for men and seven drinks for women per week) had better cognitive functioning than non-drinkers, occasional drinkers and heavy drinkers.

But there is also evidence that even moderate drinking may impair brain plasticity and cell production. Researchers in the United States gave rats alcohol over a two-week period, to raise their alcohol blood concentration to about 0.08. While this level did not impair the rats' motor skills or short-term learning, it impacted the brain's ability to produce and retain new cells, reducing new brain cell production by almost 40 per cent. Therefore, we need to protect our brains as best we can.

Excessive alcohol undoubtedly damages brain cells and brain function. Heavy consumption over long periods can damage the connections between brain cells, even if the cells are not killed. It can also affect the way your body functions. Long-term drinking can cause brain atrophy or shrinkage, as seen in brain diseases such as stroke and Alzheimer's disease.

There is debate about whether permanent brain damage is caused directly or indirectly.

We know, for example, that severe alcoholic liver disease has an indirect effect on the brain. When the liver is damaged, it's no longer effective at processing toxins to make them harmless. As a result, poisonous toxins reach the brain, and may cause hepatic encephalopathy (decline in brain function). This can result in changes to cognition and personality, sleep disruption and even coma and death.

Alcoholism is also associated with nutritional and absorptive deficiencies. A lack of vitamin B1 (thiamine) causes brain disorders called Wernicke's encephalopathy (which manifests in confusion, unsteadiness, paralysis of eye movements) and Korsakoff's syndrome (where patients lose their short-term memory and coordination).

So, how much alcohol is okay?

To reduce the lifetime risk of harm from alcohol-related disease or injury, the National Health and Medical Research Council recommends healthy adults drink no more than two standard drinks on any day. Drinking less frequently (such as weekly rather than daily) and drinking less on each occasion will reduce your lifetime risk.

To avoid alcohol-related injuries, adults shouldn't drink more than four standard drinks on a single occasion. This applies to both sexes, because while women become intoxicated with less alcohol, men tend to take more risks and experience more harmful effects.

For pregnant women and young people under the age of 18, the guidelines say not drinking is the safest option.

So while alcohol may not kill brain cells, if this myth encourages us to rethink that third beer or glass of wine, I won't mind if it hangs around.

You can't mix antibiotics
with alcohol

Michael Vagg, *Clinical Senior Lecturer, Deakin University School of Medicine; and Pain Specialist, Barwon Health*

Staying off alcohol when taking antibiotics has been hallowed advice from GPs, pharmacists and well-meaning relatives for decades.

It's difficult to work out exactly where the advice originated, but Karl Kruszelnicki (Dr Karl) suggests it dates back as far back as the 1950s, when penicillin came into use as the first really effective treatment for sexually transmitted infections (STIs) such as gonorrhoea and syphilis.

Doctors were apparently worried that disinhibited acts under the influence of the demon drink could undo their expensive treatment with the new miracle drugs. So patients were advised to abstain (from alcohol) until things cleared up.

Survey evidence suggests these fears may be well founded. Participants receiving treatment for STIs at a United Kingdom clinic were more likely to engage in risky sexual activity while intoxicated.

The advice that you shouldn't drink alcohol while taking antibiotics does hold true for a small group of anti-infective drugs including metronidazole (Flagyl, Metronide or Metrogyl), tinidazole (Fasigyn or Simplotan) and sulfamethoxazole/trimethoprim (Bactrim, Co-trimoxazole). These drugs block one of the major pathways that metabolise alcohol and cause a rapid build up of nasties called acetaldehydes, which are responsible for many of the unpleasant physical effects of hangovers. With these drugs on board, you can be red-faced, fainting and vomiting after as little as one glass of beer.

But these anti-infective drugs have fairly specialised uses – to treat infections with organisms such as giardia (from contaminated drinking water) or intestinal worms, for instance – and it would be unusual to be prescribed these drugs without a long lecture from your doctor or pharmacist about the potential adverse reaction.

For nearly all other types of antibiotics there is no clear evidence of harm from modest alcohol intake. You can find a comprehensive but readable summary of alcohol and medication interference at http://pubs.niaaa.nih.gov/publications/arh23-1/40-54.pdf.

But this doesn't mean it's a good idea to drink to excess when you're in the grip of an infection, as the sedative and nauseating effects of the alcohol are likely to increase if you are unwell.

Alcohol-induced dilation of blood vessels in the limbs interferes with your body's attempts to raise a fever to slow the spread of infection. Your kidneys will be forced by the alcohol to lose more fluid, thus increasing the risk of dehydration. And the deep, aching muscle pain produced by viral infections may be more likely to lead to serious muscle damage when combined with binge drinking.

Some antibiotics such as isoniazid and flucloxacillin (Flopen, Staphylex) may inflame the liver (causing mild hepatitis) in a small percentage of those treated. A boozy night out could further irritate the liver, which is already working hard to get rid of the extra alcohol. A similar mild hepatitis may occur with some infections such as glandular fever, which would have the same outcome.

So if you're unwell and thinking of having a big night out, it's better to go easy on the alcohol whether you are on antibiotics or not. You'll recover quicker and you'll reduce your risk of secondary complications.

If you're on one of the problematic drugs, it's important to take the 'no alcohol' warning seriously or you'll quickly and deeply regret even a few mouthfuls of alcohol.

For most antibiotic users, though, a glass of bubbly or a cold beer should be fine.

A diet high in antioxidants slows the ageing process

Michael Vagg, *Clinical Senior Lecturer, Deakin University School of Medicine; and Pain Specialist, Barwon Health*

As Australians' life expectancy nudges past 80 years, it's no surprise that we're searching for ways to add youthfulness and vitality to our later years.

It's a nice idea that a good dose of blueberries, pomegranates, green tea or even an antioxidant supplement could reduce the impact of ageing on the body. But does the science behind antioxidants stack up?

One of the most persuasive scientific ideas in the field is the Mitochondrial Free Radical Theory of Ageing (MFRTA), first proposed back in the 1950s, which explains the ageing process as the result of 'oxidative stress'.

The chemistry is complex but it boils down to the idea that free radicals (ions with an unbalanced charge) react very readily with biological molecules and are subsequently damaged. The changes of ageing – organ deterioration, sagging skin, poorer healing, and so on – are therefore due to accumulated injury.

This theory is supported by some laboratory results, such as the observation that an animal's life span roughly correlates with its metabolic rate (ability to expend energy) and amount of antioxidant activity in a species.

Studies have also shown that raising antioxidant levels in animals seems to increase their life span.

Limiting calorie intake has been found to reduce the production of Reactive Oxygen Species (ROS), which are the most common free radicals in the body.

It's about at this point that the hype begins to take over from the science.

As scientists have interrogated the MFRTA over the past decade, the results have shown some gaping holes in the theory.

More detailed animal research suggests that the longest living animals have low levels of ROS damage simply because they produce lower levels of these free radicals. In fact, the entire relationship of oxidative damage to longevity has also been disputed.

But it's possible to be right for the wrong reason in science, as ideas are constantly being examined, refined and discarded.

So what about the results of dietary antioxidants in the real world?

Again, the relationship is very complex – early studies linking high dietary antioxidant intake with improved health and longevity have not been reproduced.

In fact, the rates of some types of cancer may be increased in people consuming high amounts of antioxidants.

Perhaps most worrying for people who hit the gym is that antioxidant supplements may reduce the effectiveness of exercise training by preventing the muscles from adapting as well to the effects of the training.

So a fair summary of the science is that while MFRTA has been a useful and productive scientific hypothesis, it's unlikely to be true in its pure form.

That means for the time being, you can afford to leave the expensive antioxidant supplements on the shelf and choose foods based on their nutritional value.

As for the recipe for youthfulness and vitality in older age, good genes and regular exercise seems to be the best combination.

Take an aspirin a day after you turn 50

Michael Tam, *Lecturer in Primary Care, University of NSW*

Aspirin is a historical marvel. It's been manufactured for more than a century and is still in widespread use. No other medication can claim as many different narratives and uses as this analgesic. It's been known as:

- a traditional medicine – aspirin-like treatments, based on salicylate, have been derived from plants such as willows for millennia;
- an international blockbuster – at the turn of the 20th century, aspirin was one of the few effective treatments for fever and pain, and was wildly popular (and profitable);
- a hazard to children – aspirin was recognised in the 1980s as a potential cause of childhood death (Reyes Syndrome, associated with aspirin use to treat a viral illness); and
- a modern wonder-drug – aspirin has been resurrected as an important and inexpensive medication for the prevention and treatment of heart attacks and strokes.

And there are many fascinating tales of intrigue, international politics and corporate espionage in aspirin's history.

German affiliates undermined the manufacture of explosives in the United States during World War I by cornering the market of a key ingredient, under the guise of aspirin production. And Germany was forced to hand over the trademark 'Aspirin' as part of war reparations in the Treaty of Versailles.

In the modern context, it is commonly believed that once individuals reach a certain age, it's wise to take 'an aspirin a day' for good health.

This narrative starts in 1948 with Dr Lawrence Craven, a general practitioner, in California. He had observed that aspirin was a mild blood thinner and reasoned that it might be able to prevent heart attacks.

Dr Craven enrolled his male patients, aged 40 to 65, into a clinical trial and asked them to take aspirin daily. In the 1950s, he published three articles on his trial and concluded that aspirin appeared to protect his patients from heart attacks and strokes.

Dr Craven died in 1957 (of a heart attack!) and his results – which were published in the obscure *Mississippi Valley Medical Journal* – were promptly forgotten.

How aspirin works in clotting and bleeding was discovered in the 1960s. And by the '70s and '80s, aspirin was tested in clinical trials for heart attacks and strokes. These studies demonstrated that aspirin was effective in preventing further heart attacks or strokes (known as secondary prevention).

In the 1990s, our 'medical myth' was not considered a myth. The American College of Chest Physicians (ACCP), a respected group that publishes guidelines on the use of blood thinners, recommended that aspirin 'be considered for all individuals over age 50 years who are free of contraindications'.

But others were less confident about such a broad recommendation. First, although aspirin was unambiguously beneficial for those who already had cardiovascular disease, the evidence was less clear for those who did not (such use is known as primary prevention).

Second, long-term aspirin therapy has potential harms – it increases the risk of bleeding, which, in some cases, can be life-threatening. Conceptually, if an individual's risk of cardiovascular disease is low, then the potential benefit of aspirin would not outweigh the potential harms from bleeding.

The most recent recommendations from the ACCP (published February 2012) are a 'soft' suggestion for aspirin for primary prevention in those aged 50 years and above. It recognises that the benefits to heart attacks and strokes are closely matched with the risk of major bleeding.

The authors were swayed by some recent data suggesting aspirin might lower cancer risk and death. Nevertheless, they emphasised the need for shared decision-making between doctors and patients.

So, is that the end of this particular aspirin narrative?

Not quite. In keeping with the drama of the history of aspirin, a major study examining the role of aspirin in primary prevention was published in the same month as the ACCP guidelines. It confirmed that the benefits of lowering heart attacks and strokes were similar to the increased risks of bleeding.

Importantly, the study found no reduced risk of cancer, contrary to previous reports.

Behind this lack of clarity is the uncertainty of small numbers – trying to balance a small gain with a small risk. For someone who has never had a heart attack or stroke, the likelihood of benefit from aspirin is low, but the payoff could be massive. Similarly, the odds of being harmed by aspirin are also low, but could be catastrophic if it occurred.

Those aged over 50 without a history of cardiovascular disease may benefit from regular low-dose aspirin. But that depends on their individual risk (and perceptions of risk) of heart attack, stroke and major bleeding.

So before you pick up the aspirin for your daily dose, talk to your GP about the potential risks and benefits for you.

Cutting carbs is the best way to lose weight

Gary Sacks, Research Fellow, Deakin Population Health,
Deakin University

There seems to be an endless number of fad diets and 'golden rules' for weight loss. One of the most popular of these rules is that cutting carbohydrates (carbs) is the best way to lose weight.

The most famous low-carb diet is the Atkins diet, first developed in the 1970s. The Atkins diet recommends limiting foods high in carbs, such as bread, pasta and rice. Carbs are replaced with foods containing a higher percentage of proteins and fats (meat, poultry, fish, eggs and cheese) and other low-carb foods (most vegetables).

But what does the evidence show us about whether low-carb diets really are best for weight loss?

Theoretically, a 'calorie is a calorie' and it doesn't matter what types of food the calories come from. Accordingly, all reduced-energy (calorie) diets should lead to equivalent weight loss.

However, some studies have reported that low-carb diets lead to greater weight loss than other types of diets, at least in the short term. So, what are the possible explanations for these results, and can we rely on them?

1. Changes in body composition

Energy is stored in the body as protein, fat and glycogen, which is a form of carbohydrate. If there is an imbalance between how many of these nutrients are ingested (through the food that is eaten) and how many are used by the body for every day functions, body composition will change.

In turn, this will affect bodyweight because of the different impact that the relative amounts of stored protein, fat and carbohydrates have on it.

However, most of the studies in which researchers have measured calorie intake very accurately (that is, they've locked people in a room and measured exactly what they've eaten for several days), show absolutely no difference in weight loss based on the composition of the diet. High-protein diets and high-carb diets resulted in the same weight loss.

This indicates that, in the short term at least, the human body is a superb regulator of the type of energy it uses, and whether the diet is low-carb or high-carb probably won't make much of a difference to the amount of weight lost.

2. Changes in metabolic rate

The body's metabolic rate (the amount of energy expended by the body in a given time) is dependent on the composition of the diet. Consumption of protein, for example, is known to result in a larger increase in energy expenditure for several hours after a meal compared with the consumption of fat or carbs.

But the overall effect of diet composition on total energy expenditure is relatively small. As a result, the assumption that a 'calorie is a calorie' is probably a reasonable estimation as far as energy expenditure is concerned.

3. Changes in hunger levels and satiety

Some diets can lead to reduced hunger, improved satiety (feeling full), and can be easier to stick to than others. There is an enormous amount of research on this.

The problem is that it's extremely difficult to accurately measure what people are eating over extended time periods. People rarely stick to their diets for more than just a few weeks, making it almost impossible to adequately compare the effects of different diets.

And so, is cutting carbs the best way to lose weight?

Maybe, but there's not really good evidence supporting it. All diets with similar calorie content have a similar effect on weight loss in the short term. This is because the body adapts rapidly to changes in relative protein, fat and carbohydrate intake levels.

The truth is that losing weight and keeping it off in the long term is difficult. It requires permanent changes to the number of calories you eat each day.

Perhaps the best dietary advice comes from *The Omnivore's Dilemma* author Michael Pollan when he says, 'Eat food. Not too much. Mostly plants.'

Eating carrots will improve your eyesight

Harrison Weisinger, *Foundation Director of Optometry and Chair in Optometry, School of Medicine, Deakin University*

Getting enough vitamin A is important for healthy eyes. And carrots are a rich and natural source of this vitamin, which is basically a group of chemicals made up of retinal (the active form of vitamin A) and carotenes such as beta-carotene (which gives carrots their distinctive colour).

But a diet overloaded with carrots – and vitamin A – won't leave you with healthier eyes.

To understand where vitamin A fits in, I'll first explain a little about the process of vision. When we look at something, light from that object enters the eye and is focused onto the inside back surface of the eyeball, which is lined by a thin layer of cells. This is called the retina.

The retina is responsible for catching light and turning this into a neural signal, which is then sent up to the brain for further processing. In order to perform this wondrous action, the retina has specialised cells, called photoreceptors, each of which is packed with light-catching pigments.

The predominant pigment in the retina is rhodopsin, a major part of which is retinal (vitamin A). When the retinal reacts with light, it induces a cascade of biochemical events and shape changes in the rhodopsin molecule. In turn, this creates an electrical signal. This whole process is known as phototransduction, and it's really where vision begins.

Humans are unable to synthesise vitamin A afresh and, therefore, must take it in through their diet to maintain normal visual function. Vitamin A can be found in a range of meats and vegetables – the most notable being the carrot, though the best source is probably liver.

For most people living in developed countries, adequate vitamin A intake is not an issue, so eating more carrots will make no noticeable difference. This is because our diets contain enough vitamin A and we are able to store it, unlike other nutrients such as vitamin B.

In fact, well-nourished pregnant women should avoid supplementing their diet with vitamin A when their total daily intake is around 3000 IU (International Units) because too much vitamin A (well above 10 000 IU per day) can cause birth defects. But vitamin A-rich foods are safe so you can still munch on a bag of carrots without doing any harm (provided you don't mind your skin turning orange from the carotenes!).

In the developing world, however, an estimated half a million children become blind each year as a consequence of dietary vitamin A deficiency. But carrots aren't the answer, as they are not easily grown and don't last long enough to be distributed. World food programs are instead trialling vitamin A-rich bananas and sweet potatoes as a source of nutrition to improve eye health.

While there is still much to be done to prevent vitamin A deficiency in the developing world, for those reading this article, carrots will make very little difference to your eyesight.

Childhood vaccinations are dangerous

Fiona Stanley, Perinatal and Pediatric Epidemiologist; Founding Director and Patron of the Telethon Institute for Child Health Research; and Distinguished Professorial Fellow, University of Western Australia

When I was an infant I had whooping cough and was ill for three months. I don't remember it, of course, but I know it was very distressing for my parents. I do remember later trips with my researcher father to his laboratory where he worked on a vaccine for polio and to hospitals where infected children my own age were in iron lungs. That was very distressing.

I mention this because today people don't see such diseases. They aren't frightened about whooping cough or polio. In contrast, 100 per cent of parents in Western Australia had their children vaccinated against polio when the vaccine was made available in 1956. Why? They were scared of their kids getting polio, a terrible disease as reflected in its other name, infantile paralysis.

Because today's parents don't have first-hand experience with dangerous infectious diseases they can be misled by myths about the supposed dangers of childhood vaccination: for instance, that whooping cough vaccine causes brain damage; the measles, mumps and rubella (MMR) vaccine causes autism and vaccination causes cot death or sudden infant death syndrome (SIDS).

There is no truth to any of these claims. We in Australia have some of the best population data in the world on vaccination outcomes in children and it's absolutely clear these myths are just that, myths.

The whooping cough myth started in the 1974 in the United Kingdom when some parents claimed that after being vaccinated their children were diagnosed with neurological disorders, what they called 'brain damage'.

In fact, it was a coincidence. The first signs a child has a genetic or other brain disorder occur about six months of age. The vaccine is given at two, three and four months, hence the incorrect assumption that the latter caused the former.

I was a student in the UK at the time. It was disastrous that the medical and epidemiological professions didn't respond after the kids with the claims of vaccine-caused brain injury were shown on television. The government paid compensation, reinforcing the false vaccination-brain damage association.

As a result, the rate of vaccination dropped from 81 per cent to 31 per cent, triggering the most horrendous epidemic of whooping cough. In one year, 21 children died and thousands were hospitalised with severe pneumonia and, sadly, brain damage from the infection.

The fear of the disease influenced parents to vaccinate again. Immunisation rates went back up and disease incidence went down. But it's a tragedy that it took an epidemic to prove that vaccination is protective. Several major studies also demonstrated clearly that whooping cough vaccines protect against brain damage rather than cause it.

The misguided belief that vaccination causes SIDS is also a case of myth by coincidence. The peak age of SIDS is four months, following vaccinations given from two to four months. The timing of the two events is associated in people's minds, despite study after study showing no connection.

Instead, the research shows SIDS is linked strongly to lying babies on their face or having their head covered with bedding or toys. Other risk factors include smoking, not breastfeeding, overcrowding and over-heating.

The myth that the MMR vaccine causes autism is particularly naughty. It was started in 1998 by a scientist who published the claim in a widely reported paper in *The Lancet*.

Again, vaccination rates fell precipitously and outbreaks of measles, mumps and rubella occurred. It was revealed the scientist had undeclared conflicts of interest and had engaged in scientific misconduct. The paper was retracted but the damage was done.

Such myths demonstrate why it's absolutely crucial that medical researchers obtain solid laboratory data about new and combination vaccines, test them rigorously and obtain very good surveillance and monitoring data. The public must have confidence that the research is done and done well.

That's why the Australian Academy of Science has released a booklet – *The Science of Immunisation: Questions and Answers* – which explains the basics of vaccination and debunks common myths about vaccines and vaccination. It draws on expertise from a broad sector of the Australian science community, from virology and immunology to my field of epidemiology. You can download it at http://www.science.org.au/policy/immunisation.html.

I urge all Australians to get the truth about the myths. Vaccination is a wonderful development in public health. It has prevented enormous suffering and millions of deaths worldwide. The benefits of vaccination outweigh the very small risk of unwanted side effects. Just ask the parents of 1956.

Chocolate causes acne

Clare Collins, Professor in Nutrition and Dietetics, University of Newcastle

Outbreaks of pimples, blackheads and cysts are a cause of enormous anxiety and embarrassment among teens and young adults. If you're part of the 20 per cent of Australians who have experienced severe acne, you've probably tried a raft of treatments and preventive measures. But does giving up chocolate help?

It's unclear where or how this myth arose, but researchers tested the link three times from 1965 to 1971, suggesting it must have been a commonly held belief at least 40 years ago. All three studies came up with the answer: chocolate doesn't exacerbate acne.

But by today's standards, the investigations were all of a poor scientific standard. The original study, conducted in 1965, contained just eight participants.

The next study, published in the *Journal of the American Medical Association* in 1969, had 65 participants, but the results were confounded by the use of two different groups of subjects: 60 adolescents (14 girls) and 35 young adult male prisoners of an unspecified age. The researchers didn't account for the effects of gender, age, puberty, menstrual cycles, stress, smoking, lifestyle, background dietary intake or medical conditions affecting the skin.

The 1971 study, published in the journal *American Family Physician*, evaluated the effects of chocolate, milk, roasted peanuts and cola on acne in 27 students, but it failed to report their age and only followed them for one week. Also, it didn't address all potential confounders and failed to report significant acne outbreaks during or immediately after the study period.

Several detailed critiques of these studies' shortcomings have since been published. But the chocolate and acne myth has remained unchallenged.

A recent study of YouTube videos found more than 85 per cent of clips with keywords 'acne', 'acne diet' and 'acne food' supported the belief that diet has a moderate association with acne.

So why does the myth that chocolate causes acne continue to circulate?

Perhaps the fault lies with us researchers as the proponents of evidence-based practice. We have failed to subject this chocolate myth to the rigours of a randomised control trial (RCT), despite the fact that almost all people aged 15 to 17 years experience some degree of acne. We need a decent RCT so we can know once and for all whether to unleash our teenagers, and ourselves, in the confectionery aisle at the supermarket.

Food and acne

Recent evidence suggests it may be time to expand our investigation of chocolate and acne, and focus on milk consumption and the glycemic index (GI).

Milk and its products, including pasteurised milk, yoghurt, ice cream and cottage cheese, contain an array of naturally occurring ingredients that promote growth. The whey protein of dairy products, with the exception of cheese, leads to an increased release of insulin. And the casein protein in dairy products leads to an increase in levels of insulin-like growth factor (IGF).

Surprisingly, drinking milk raises blood insulin levels to a greater degree than predicted, based solely on its lactose content (the carbohydrate found in milk). Although the biochemical pathways are complex, in simple terms, this can lead to a worsening of acne.

This same reaction does not occur after eating cheese.

Glycemic index and glycemic load

Diets with a high GI or glycemic load (GL) trigger a higher insulin response. This is because high-GI foods contain carbohydrate in a form that is quickly digested and absorbed into the blood stream, sending a message to the pancreas to secrete insulin.

This high level of insulin, in turn, increases IGF, potentially exacerbating acne. The insulin then sets out to clear the glucose from the blood.

So can a high-GL diet, with more high-GI foods, help manage acne?

This was tested in a 2007 randomised control trial. The researchers asked 43 males aged 18 years to follow either a low-GL or a high-GL diet for 12 weeks. Meanwhile, the severity of their acne was assessed by dermatologists who were blinded to the dietary intervention aspects of the study.

The low-GL groups were instructed to swap some high-GI foods for others higher in protein, such as lean meat, chicken or fish, and to favour lower-GI foods such as wholegrain bread, pasta and fruits. The low-GL diet aimed for 25 per cent energy from protein, 45 per cent from low-GI carbohydrates and 30 per cent energy from fats. The high-GL group was encouraged to follow a high-carbohydrate diet.

Interestingly, those following the low-GL diet saw their acne improve, along with their insulin sensitivity. They also lost weight. It's important to note, however, that this work has not been repeated by other researchers at this stage.

What to do if you suffer from acne

Acne commonly persists into adulthood, with almost two-thirds of adults in their 20s and 43 per cent in their 30s experiencing the condition. No matter what your age, you can get help from your GP, who may need to refer you to a dermatologist.

When it comes to food, more good-quality research studies are needed to assess the impact of dietary manipulations. But along with medical treatment, there are some dietary strategies worth trying:

- Reduce your intake of high-GI carbohydrate foods, such as potatoes, doughnuts, pancakes, sweetened breakfast cereal and white bread. Swap high-GI for lower-GI choices such as apples, bananas, carrots, corn, muesli, mixed grain bread, pasta, porridge, tomato soup and sweet potato.

- Be more active to improve insulin sensitivity; go for a short walk after eating to help reduce blood sugars and moderate insulin levels.
- Reduce your milk (but not cheese) intake. To achieve peak bone mass you will need to take a daily calcium supplement.
- If you're overweight, try and reduce your weight, even by a few kilograms.

Chocolate is an aphrodisiac

Merlin Thomas, Adjunct Professor, Preventive Medicine, Baker IDI
Heart & Diabetes Institute

There are many ways to a woman's heart. But is a box of chocolates really one of them?

What makes chocolate romantic is entirely contextual. Valentine's Day is traditionally the time for couples to profess their love for one another, usually by giving chocolate or flowers and sending greeting cards or, now, e-valentines. Chocolate Easter eggs hold no such allure.

But if its role in romance is just symbolism, why should chocolate take the cake?

One reason may be that cocoa products have historically been considered an exclusive item; the Aztecs thought it was their gods' drink of choice. The scientific name of the cocoa tree, *Theobroma cacao*, actually comes from the Greek words *theo* (god) and *broma* (drink).

So, if you worship your lover and think her a goddess, isn't chocolate a fitting tribute?

Cocoa products contain many biologically active components (including methylxanthines, biogenic amines, flavanols and cannabinoid-like fatty acids) that could, in theory, impact on human health. Some studies suggest regular chocolate intake is associated with a reduced risk of cardiovascular disease and mood disorders.

But what we really want to know is whether chocolate is an aphrodisiac.

It's tempting to hypothesise that chocolate has a direct impact on female sexuality – it's almost believable. Some studies have even suggested women who eat chocolate have stronger libidos than those who don't. But this is not the same as cause and effect.

There's no biological evidence to show that chocolate – or any other food or beverage – works as an aphrodisiac. Several foods have been ascribed aphrodisiac qualities, and they tend to have a strong placebo effect. In other words, they get you thinking about sex, and this puts sex on your mind.

Many products have acquired aphrodisiac reputations simply because they were once exotic or unfamiliar foods. Before Hershey, globalisation and mass production, chocolate was the inaccessible luxury of the rich and famous. Who wouldn't want a bite of that?

Another part of chocolate's aura is sympathetic magic. This idea posits that if two things are alike, then it might be possible to garner the same effects from them. This is also known as the law of similarity and explains the fallacious appeal of rhino horn!

But sex and chocolate have much in common. Both cause blood vessels to dilate (known as vasodilatation) and accentuate flushing, especially as chocolate was traditionally a hot beverage.

Finally, when it comes to hedonistic appeal, the taste, texture, aroma and packaging of chocolate are hard to beat. The sensory qualities of creamy chocolate melting in your mouth may be far more stimulating to the brain than the same chocolate in your stomach.

In mammals, taste and odour are among the most important determinants of sexual attractiveness. The existence of an equivalent human pheromone remains to be established. But if there were one, it would probably smell and taste like chocolate on Valentine's Day.

Cranberry juice prevents bladder infections

Michael Tam, Lecturer in Primary Care, University of NSW

You might eat them in a sauce alongside your Christmas turkey or drink them juiced, perhaps with a shot of vodka. But the sweet, tart cranberry is also well known as a remedy for preventing urinary tract infections (UTIs).

Cystitis – an infection and inflammation of the lining of the bladder – is the most common form of UTI, with symptoms including:

- the frequent urge to pass urine;
- a stinging or burning sensation when passing urine;
- smelly urine;
- cloudy or bloody urine; and
- pain in the low abdomen or pelvis.

This condition occurs frequently in women, with one in three experiencing cystitis at least once. As a general practitioner, it would be unusual for me to not see a case of cystitis most weeks. In most cases, cystitis is easily treated with a course of antibiotics.

As a folk remedy with a long history among Native Americans, cranberry juice was dismissed for years by the medical establishment. But this changed in the 1980s and 1990s when it was discovered that cranberry juice contained chemicals that seemed to stop *E. coli* (the most common bacteria causing UTIs) from sticking to the lining of the bladder.

In theory, if bacteria cannot attach to the bladder lining, then they would be flushed out with the urine and thus not cause an infection.

This thinking has been popularised in the last couple of decades. Cranberry juice and capsules have been widely recommended and promoted as a treatment for preventing bladder infections, particularly for women who suffer from recurrent infections. Health literature aimed at consumers, including high-quality sources, often advises that cranberry products can be used to reduce the frequency of UTI episodes.

Given this, it would be natural to believe that cranberry products were a proven therapy! Indeed, I was taught in medical school that cranberry was effective, and have personally prescribed it for my patients in the past.

Curiously, although there appear to be good scientific reasons why cranberry products could work in preventing UTIs, evidence for this in real patients has been rather murky.

A 2009 Cochrane Library systematic review, which independently analysed all the available evidence, noted there was some evidence that cranberry products *might* work, but it wasn't clear what the 'optimum dosage or method or administration' was.

The large number of dropouts from the available trials also suggested that it might not have been an acceptable treatment over a longer period of time.

This review was updated in October 2012 with the inclusion of newer and larger studies. Disappointingly, this revised appraisal of the empirical evidence seems to suggest that cranberry does not reduce the likelihood of a recurrence of UTIs in women.

I doubt that we have heard the last word on cranberry and there are studies in the pipeline.

But the weight of evidence, especially those from larger and better-designed trials, points towards the likelihood that cranberry products are ineffective for preventing UTIs.

Take a vitamin a day for better health

Clare Collins, *Professor in Nutrition and Dietetics,*
University of Newcastle

Forget an apple a day, vitamin manufacturers would have you believe it's important to take daily vitamins to boost your health.

And a surprising proportion of Australians do. Data from the last National Health survey (back in 1995) showed that up to 30 per cent of Australians had recently taken vitamin or mineral supplements – mostly for preventive health reasons.

More recently, the 45 and Up study of more than 100 000 Australian adults found that 19 per cent of men and 29 per cent of women reported taking vitamin or mineral supplements.

But most healthy people don't need to take vitamins. A better safeguard for your health would be to spend the money you save from *not* buying supplements, on buying more vegetables and fruit.

The *Australian Guide to Healthy Eating* (AGHE) translates the national dietary guidelines into recommended daily food serves to help Australians eat better, without the need for vitamins or mineral supplements. You can view it at http://www.eatfor-health.gov.au/guidelines/australian-guide-healthy-eating.

In a nutshell, the aim is for adults to have a minimum daily intake of:

- two serves of fruit;
- four to five serves of vegetables;
- four to six serves of wholemeal or wholegrain breads and cereals;
- two serves of reduced fat dairy products;

- one serve of lean protein; and
- a small amount of healthy fats.

The problem is, we just don't follow the advice in the dietary guidelines, or eat like the patterns suggested in the AGHE.

The last National Nutrition Survey of dietary intakes in adults (from 1995 – this is currently being updated) found that we had inadequate intakes of vegetables, fruit, wholegrain cereals and dairy products. We also consumed too much fat, especially saturated fat, and over a third of our daily energy intake came from energy-dense nutrient-poor foods – 'junk' foods.

So what do we do: turn to vitamin and mineral supplements to make up the shortfall? Or try harder to encourage Australians to eat better?

I vote for the second approach because taking supplements is not without risks.

Take lung cancer, for example. Epidemiological research indicated that eating more fruit and vegetables was associated with a reduced risk of lung cancer. After this relationship was recognised, several clinical trials then gave people supplements of beta-carotene, because it's a major carotenoid (pigment) in vegetables and fruit.

But the supplements had the opposite effect and actually increased the risk of lung cancer in smokers.

Medical problems that arise due to excessive intakes of vitamins and minerals are almost always due to intakes of supplements. To develop toxicity from vitamins in food you'd have to eat excessive amounts of specific foods such as carrots (which could make your skin turn yellow) or liver (vitamin A toxicity would leave you with blurred vision, dizziness, nausea and headaches).

There are, however, people with health conditions or in a particular life stage when they really need vitamins. This includes people with chronic medical problems (such as cystic fibrosis, coeliac disease, pancreatitis), people on restrictive diets to achieve rapid weight loss, and those with conditions that interfere with their ability to eat properly.

Women planning a pregnancy also require additional nutrients. Folic acid supplements are strongly recommended in early pregnancy to reduce the risk of having a baby with neural-tube defects such as spina bifida.

Let's leave vitamin supplements to those who need them, and call this myth busted.

Dairy products exacerbate asthma

Janet Rimmer, *Respiratory Physician and Allergist; and Clinical Associate Professor, Sydney Medical School, University of Sydney*

Dairy products are good for the bones, so we're encouraged to have regular serves of (reduced-fat) milk, cheese and yogurt. But can they make asthma and allergies worse?

Asthma is a respiratory condition that causes the airways to the lungs to constrict when exposed to certain triggers, making it difficult to breathe. One in 10 Australian adults and about one in nine children will suffer from asthma during their lifetime.

People with asthma generally aren't put on a restrictive diet because it's rare that food allergens trigger the illness. It's more likely that food additives or food preservatives such as sulphur dioxide (identified on food labels by the number E220) will trigger asthma. This is relevant for 5 to 10 per cent of asthmatics who may need to avoid the additive.

Specific cow's milk-related diseases include cow's milk allergy, food protein-induced enterocolitis (FPIES), lactase deficiency (or lactose intolerance) and milk intolerance.

Around 2 per cent of babies are allergic to cow's milk. In this group, the ingestion of dairy products *can* cause asthma as well as other problems such as hives and vomiting. It's important that parents obtain a correct diagnosis for children with the condition, using skin testing or blood tests to show the presence of allergy (IgE) antibodies to milk.

About 80 per cent of children will grow out of their cow's milk allergy. But while the allergy persists, it's important to seek medical advice about alternative sources of nutrition and when to consider re-introducing milk.

In the other cow's milk-related diseases such as FPIES, lactose intolerance and milk intolerance, the ingestion of milk will cause symptoms – usually gastrointestinal, such as diarrhoea and vomiting – but will not aggravate asthma.

Respiratory allergies such as asthma and rhinitis (hay fever) are usually triggered by what we inhale rather than what we eat.

Some people complain that the ingestion of milk causes a runny nose, makes their throat feel as though it is coated by thick mucus and triggers coughing. But research studies have shown that these sensations are due to the texture of the milk and can be similarly caused by fluids of the same thickness.

Studies have also shown that in people with asthma, the ingestion of milk has no effect on lung capacity and does not trigger asthma symptoms. Drinking cold milk may cause a cough in patients with asthma but this is more likely to be due to the temperature of the milk and can be avoided by warming the milk.

In children with asthma, a runny nose is more likely to be due to associated allergic rhinitis or a viral infection rather than milk in the diet.

Why we need calcium

Calcium is vitally important in the body for cell functioning. It's stored in the bones and teeth where it supports their structure and function. Dairy products are the main source of calcium in our diet.

There are particular times in life – such as during growth spurts in children and adolescents – when new bone formation occurs and adequate dietary calcium is essential to facilitate this process. Maximum bone density is achieved during puberty and the higher it is at this time, the better one's lifetime bone health will be. This is why adequate dietary calcium intake is especially important in childhood.

Further, many asthmatics are prescribed preventer medication which contains an anti-inflammatory corticosteroid medicine. At high doses, this is associated with the development of osteopenia (a precursor to osteoporosis) and osteoporosis itself.

So it's important that asthmatics of all ages have an adequate calcium intake to meet their dietary needs:

- 210 mg daily from birth to six months;
- 500 mg in early childhood;
- 1300 mg from ages 12 to 18 years; and
- 1000 to 1200 mg in adults.

These targets are difficult to achieve unless dairy products (milk, yoghurt, cheese) are part of the diet. If you're not getting enough calcium, talk to your doctor or health professional about calcium supplements such as calcium carbonate or calcium citrate.

Deodorants cause breast cancer

Terry Slevin, *Honorary Senior Lecturer in Public Health, Curtin University; Education & Research Director, Cancer Council WA; and Chair National Skin Cancer Committee, Cancer Council Australia*

The fear that using deodorants and antiperspirants might increase the risk of breast cancer has been around for at least 15 years, probably longer.

The theory suggests that either parabens (preservatives previously used in some deodorants that act as a weak form of oestrogen) or the aluminium salts used in many antiperspirants, enter the body and contribute to or cause breast cancer.

Studies detecting the presence of traces of paraben and aluminium products in breast tissue and breast tumour are put forward as evidence of the connection.

The other argument supporting this theory centres on the high proportion of breast cancer lesions that are located in the upper outer quadrant of the breast. This is where deodorants and antiperspirants come into most contact with breast tissue.

Others have observed there is simply more breast tissue in that part of the breast. So if lesions are evenly spread, we would expect to find more disease in that part of the breast.

Another issue is measurement precision. As reported in the study that advanced the theory, between 1980 and 1996 cancer lesion locations were not always reported by breast quadrant. Only 17.5 per cent of cases recorded cancer location by quadrant, making meaningful analysis difficult.

While parabens were found in breast tissue or breast lesions, detectable measures in tissue does not in itself prove causation of disease. Breast cancers, like most solid tumours, develop their own access to the body's blood as a means to grow. As a result, it's likely that any substance that's in the bloodstream will be detectable in small amounts in the tumour tissue. But it doesn't mean the detected substance caused the cancer.

Nonetheless, as a result of the stories circulating about the potential harms of parabens, most deodorant manufacturers have stopped using these preservatives. Not because of a proven harm, but because of a suspicion ('market perception') of possible harm, which ultimately affects sales.

What does the evidence say?

Studies aimed at determining if a connection between underarm products and breast cancer really exists have not been able to find a causal link. One study in 2002 looked at about 800 women with breast cancer and a similar number of matched controls. Researchers asked about the use of antiperspirants and deodorants, and underarm shaving habits. They could not find any difference in these behaviours between those with and those without breast cancer.

Another small case control study in 2006 found that 82 per cent of the controls (women without breast cancer) and 52 per cent of cases (women with breast cancer) used antiperspirants, indicating that using the under arm product might protect against breast cancer. While the study is too small to justifiably make such a claim, it certainly does not support the 'antiperspirants cause cancer' story.

Reputable groups like the American National Cancer Institute, Cancer Research UK, the American Cancer Society and most other major authorities suggest the link between deodorant or antiperspirant use and breast cancer is unconfirmed, or simply a myth.

What about radiotherapy?

Another factor that perpetuates this myth is that patients undergoing radiotherapy are commonly advised to stop using antiperspirants during therapy, on the theory that the aluminium salts may influence the therapy. However, a 2009 Australian study indicated that less than half of patients complied with this advice, with many forgetting (43 per cent) or ignoring (10 per cent) it.

Interestingly, this study also found '[of] the 233 women who routinely wore a deodorant but abstained during radiotherapy, 19 per cent expressed a lot of concern about body odour and 45 per cent were slightly concerned'. This suggests that many people see a clear benefit in using these products.

Even more recently, a Canadian study found no evidence of antiperspirant use having any adverse effect on radiotherapy treatment for breast cancer.

It's impossible to ignore that the majority of research on the possible link between underarm cosmetics and breast cancer comes from one research group. And it seems that despite the absence of evidence to support the link, their search to prove the theory is unlikely to stop.

Who knows? They may ultimately be proven to have been correct. But based on the evidence from most of the other groups researching this question – it seems likely to remain nothing more than a myth.

Detox diets cleanse your body

Tim Crowe, *Associate Professor in Nutrition, Deakin University*

Detox diets make amazing promises of dramatic weight loss and more energy – all achieved by flushing toxins from the body. Toxins have very little to do with it; detox diets 'work' because of the very severe dietary and energy restrictions they require people to follow.

Detox or liver-cleansing diets have been around for many years. With amazing claims of rapid and easy weight loss and improved health, together with a heavy dose of Hollywood celebrity endorsement, it is no wonder these diets are in the public spotlight.

Toxin build-up from our environment and poor diet and lifestyle habits are claimed to be the main culprits for weight gain, constipation, bloating, flatulence, poor digestion, heartburn, diarrhoea, lack of energy and fatigue. 'Detoxing' is a way for the body to eliminate these toxins and as a result, a person will feel healthier and lose weight.

Detox diets can vary from a simple plan of raw vegetables and unprocessed foods and the elimination of caffeine, alcohol and refined sugars to a much stricter diet – bordering on starvation – with only juices consumed.

Some detox programs may also recommend vitamins, minerals and herbal supplements. Detox diet programs can last anywhere from a day or two to several months.

Do detox diets work?
There is no shortage of glowing testimonials from people who have gone on a detox diet, claiming to feel cleansed, energised and

healthier. Promoters of detox diets have never put forward any evidence to show that such diets help remove toxins from the body any faster than our body normally eliminates them.

The idea that we need to follow a special diet to help our body eliminate toxins is not supported by medical science. Healthy adults have a wonderful system for removal of waste products and toxins from the body. Our lungs, kidneys, liver, gastrointestinal tract and immune system are all primed to remove or neutralise toxic substances within hours of eating them.

As for the dramatic weight loss typically seen, this is easily explained by the very restrictive nature of detox diets, which can cut kilojoules dramatically.

Claims made that the typical physical side effects such as bad breath, fatigue and various aches and pains are evidence that the body is getting rid of toxins just do not stand up to scientific scrutiny. Bad breath and fatigue are simply symptoms of the body having gone into starvation mode.

The many downsides of detox diets

Apart from the false claim that a detox diet is actually 'detoxifying' the body, these diets have many well-documented downsides, including:

- feelings of tiredness and lack of energy;
- cost of the detox kit if a commercial program is followed;
- expense of buying organic food if required;
- purchasing of supplements if recommended by the diet;
- stomach and bowel upsets; and
- difficulties eating out and socialising, as most restaurants and social occasions do not involve detox-friendly meals.

The biggest downside of detox diets, especially the more extreme ones, is that any weight loss achieved is usually temporary and is more the result of a loss of water and glycogen (the body's store of carbohydrate) instead of body fat. This means that the weight lost is easily and rapidly regained once the person reverts to a more normal eating plan. These dramatic weight fluctuations can be demoralising and lead to yo-yo dieting.

Following a typical detox diet for a few days has few real health risks in otherwise healthy individuals. Very restrictive detox diets, such as water- or juice-only fasting, can be an unsafe form of weight loss and should not be used for more than a few days.

The verdict of *Choice*

Consumer magazine *Choice* has carried out several surveys and expert reviews of popular detox diets sold in supermarkets and chemists.

Choice found no sound evidence that we need to 'detox', or that following a detox program will increase the elimination of toxins from your body. Some of the popular detox kits have diet plans that are far too restrictive, and give dietary advice with either poor or no rationale.

Detox diets may do little harm to most people, except perhaps for their bank balance, but neither do they do a lot of good just on their own. Concerted changes to diet and lifestyle habits are far more valuable than detox diets and supplements.

Reading in dim light ruins your eyesight

Harrison Weisinger, *Foundation Director of Optometry and Chair in Optometry, School of Medicine, Deakin University*

The idea that reading in dim light ruins your eyes isn't my favourite old wives' tale about 'leisure activities' causing blindness, nor is it the most obscene! In any case, it's simply not true.

I'll begin my exposé with a brief explanation of how we see.

The eyes are equipped to catch the light reflected off, or generated by, objects in our world. When light enters our eyes, it's focused by the front layers of the eye onto the retina – a delicate layer of just a few rows of cells, less than half a millimetre thick – which sits at the back of the eye.

The light that falls on the retina is referred to as the image. The retina begins the process of decoding the image and sorting this into information that tells our brains about its brightness, colour, shape, size and movement. The information, in the form of neural signals, is then passed back to the brain, which further processes the data before bringing it to the attention of our conscious mind.

Some animals, such as owls, have retinas that are specialised for seeing even the tiniest amount of light. Put simply, they see clearly in conditions that we consider pitch black because their retinas contain 'rod' detectors. These rods are very sensitive, but can't decode colour (owls are colour blind). There are close to 60 000 of these rods per square millimetre of retina, which translates to owls' incredible sensitivity and sharpness of vision – called acuity.

Humans also have many rods, which is why we're able to see when driving at night when there's very little light around. But rods become useless in bright or even normal light levels. That's why we also have 'cones', which are much better in daylight and allow us to see colour. These cones can be found throughout the retina, with the greatest number in the centre.

When you turn off the lights, your 'night vision' gradually kicks in over six or seven minutes, as you stop using your cones and start using your rods.

There are no rods at the centre of the human retina, which is why we have low sensitivity to dim light in our central vision. That's the reason that, when you go outside on a clear night and look directly up at a faraway star, you won't be able to see it. You can only see a star by looking to the side of it, thus using your rods.

On the flipside, in order to read, we in practice only use the cones in the centre of our vision. To test this for yourself, try reading a column on the screen while looking just to the side of that column – impossible.

So reading in dim light – or reading at all – is possible when there is enough light around for the cones to pick up a signal.

Your eyes won't be harmed but you may give yourself a headache. This is because, from an evolutionary perspective, the eyes weren't designed for straining to see close-up objects for sustained periods. They are much better suited for looking out into the distance over fields of buffalos (not that many of us have that luxury in our modern-day lives).

Of course, eyestrain – the feeling of tired or aching eyes and headache – may indicate that you need glasses, or perhaps the glasses you're wearing may need an update. If you are concerned about the health of your eyes, see your optometrist for a check up.

Eat for two during pregnancy

Susie de Jersey, *Senior Dietitian-Nutritionist, Royal Brisbane and Women's Hospital; and Visiting Research Fellow, School of Exercise and Nutrition Sciences, Queensland University of Technology*

We've all heard people spout the phrase, 'Go on, you're eating for two now' at barbecues, dinner parties and wherever food is being served, forcing pregnant women to decline offers of more and more food from well-meaning friends and family.

While pregnant women *don't* have to eat twice as much food, the growth and development of a baby certainly does rely heavily on its mother's nutrient stores and intake during pregnancy. The Dutch Famine in 1944–45 demonstrated that poorly nourished mothers were more likely to give birth to babies with restricted growth. Their children were also more susceptible to chronic diseases in adulthood.

During pregnancy, a woman's nutrient requirements increase by between 10 and 50 per cent, depending on the specific nutrient. But her energy intake only needs to increase in the range of 15 to 25 per cent. In Western societies, excess energy and bodyweight are more common than nutritional inadequacies.

The amount of food a woman consumes during pregnancy shouldn't increase substantially. Generally, it should only increase by the daily equivalent of two medium-sized pieces of fruit and half a glass of reduced-fat milk averaged over the pregnancy term. But everyone is different.

If women do kick back and eat for two, they're likely to gain too much weight, particularly if there's not a substantial increase in physical activity. My colleagues and I have found that one-third of Australian women who were a healthy weight and just over half of women in the heavier-than-healthy category gained too much weight during their pregnancy.

The complications arising from gaining too much weight during pregnancy include a greater risk of developing gestational diabetes, problems during labour for both mother and baby, weight retention after delivery for mothers and an increased likelihood that the child will become overweight later in life.

This is not to say weight gain should be unduly restricted. Not gaining enough weight can have negative consequences for both mother and baby, so it's important to achieve a healthy balance.

There are several resources available to guide a healthy weight gain in pregnancy. A doctor, nurse or dietitian will be able to provide information specific to each woman in pregnancy, but here is a starting guide, based on pre-pregnancy body mass index (BMI):

- Women who are underweight (with a pre-pregnancy BMI of less than 18.5) should gain around 12.5 to 18 kg.
- Women of a healthy weight (pre-pregnancy BMI 18.5 to 24.9) should gain around 11.5 to 16 kg.
- Women who are overweight (pre-pregnancy BMI 25 to 29.9) should gain around 7 to 11.5 kg.
- Women who are classified as obese (pre-pregnancy BMI above 30) should gain around 5 to 9 kg.

So how do women meet the extra nutrient needs without piling on the kilos?

A pregnant body becomes more efficient at absorbing nutrients. A high-quality diet is still important but there is not as much room for those discretionary foods that have few nutrients but loads of energy.

It's important to eat two serves of fruit and five serves of vegetables every day. Lean meat, reduced-fat dairy products, and wholegrain breads and cereal products will ensure women get plenty of nutrients without overdoing the kilojoules.

Physical activity is also important – maintaining an active lifestyle and getting 30 minutes of physical activity each day will help achieve a healthy weight.

It's difficult to meet folic acid and iodine requirements during pregnancy through a regular diet. So folic acid and iodine supplements are now routinely recommended for at least one month before an intended pregnancy and for the first trimester. Ideally, iodine supplementation should continue during pregnancy and breastfeeding.

There is insufficient evidence to support taking other vitamins or a multivitamin unless low levels are diagnosed.

While it might be nice to indulge during pregnancy, the 'eating-for-two' myth should be discarded to give babies the best chance of optimal development and future health.

You need eight hours of continuous sleep each night

Leon Lack, *Professor of Psychology, Flinders University*

We're often told by the popular press and well-meaning family and friends that, for good health, we should fall asleep quickly and sleep solidly for about eight hours – otherwise we're at risk of physical and psychological ill health.

There is some evidence to suggest that those who consistently restrict their sleep to less than six hours may have increased risk of cardiovascular disease, obesity and diabetes. The biggest health risk of sleep deprivation comes from accidents, especially falling asleep while driving.

Sleep need varies depending on the individual and can be anywhere from 12 hours in long-sleeping children, to six hours in short-sleeping healthy older adults. But despite the prevailing belief, normal sleep is not a long, deep valley of unconsciousness.

The sleep period is made up of 90-minute cycles. Waking up between these sleep cycles is a normal part of the sleep pattern and becomes more common as we get older.

It's time to set the record straight about the myth of continuous sleep – and hopefully alleviate some of the anxiety that comes from lying in bed awake at night.

So what are the alternatives to continuous sleep?

The siesta
The siesta sleep quota is made up of a one- to two-hour sleep in the early afternoon and a longer period of five to six hours late in the

night. Like mammals and birds, humans tend to be most active around dawn and dusk and less active in the middle of the day.

It's thought the siesta was the dominant sleep pattern before the industrial revolution required people to be continuously awake across the day to serve the sleepless industrial machine. It's still common in rural communities around the world, not just in Mediterranean or Latin American cultures.

Our siesta tendency or post-lunch decline of alertness still occurs in those who never take afternoon naps. And this has less to do with overindulging at lunchtime and more to do with our circadian rhythms, which control our body clock, hormone production, temperature and digestive function over a 24-hour period.

Bi-phasic sleep

Historical records also suggest that a segmented or bi-phasic sleep pattern was the norm before the industrial revolution. This pattern consists of an initial sleep of about four and a half hours (three sleep cycles of 90 minutes each) followed by one to two hours awake and then a second sleep period of another three hours (another two sleep cycles).

During the winter months, northern Europeans would spend nine or 10 hours in bed, with two to three hours of it spent awake, either in one long mid-night period or several shorter wake periods across the night.

The bed was the cheapest place to keep warm and was considered a place of rest as well as sleep. A few hours of wakefulness certainly wouldn't have been considered abnormal or labelled as insomnia.

Can't sleep? Don't worry

These days we expect to have close to 100 per cent of our time in bed asleep, dozing off within minutes and not waking at all until the alarm sounds. Unfortunately this myth sets us up for worry if we find ourselves awake in the middle of the night. And this worry can lead gradually to the development of insomnia.

Humans can sleep on very different schedules, with little difference in wakeful competence. International sleep researchers have trialled several different sleep schedules: sleep for 20 minutes every hour; one hour's sleep every three hours; 10 hours' sleep every 28 hours. Participants survive easily on all these schedules despite their impracticality in our 24-hour world.

The best quality sleep is obtained during our circadian low phase – when body temperature and metabolic rate are at their lowest. For most people, this occurs late at night. But just like other species, humans can be opportunistic sleepers and satisfy our need for sleep when we get the opportunity.

There's no doubt that the eight-hour solid sleep myth is a relatively recent cultural imposition. And although it satisfies our modern lifestyle, it does have its disadvantages.

Some have lamented the loss of wakefulness between sleep cycles, seeing them as a valuable time of contemplation or creativity.

But probably the greatest negative impact of the eight-hour sleep myth is its power to create insomniacs out of good sleepers who experience normal awakenings across the night.

We're not getting enough sun

Ian Olver, *Clinical Professor of Oncology, Cancer Council Australia*

Myths abound about UV radiation and its effect on our health. We hear that sun-protection has triggered an epidemic of vitamin D deficiency; being tanned protects you from sunburn; a tan looks healthy; and 'old' skin doesn't need to be protected from the sun like 'young' skin does.

Myth, myth, myth, myth.

What is beyond doubt is that Australia is the world's skin cancer capital, yet skin cancer is the most preventable of all common cancer types.

There were 11 000 cases of melanoma diagnosed in Australia in 2008. Deaths from melanoma and non-melanoma skin cancers combined in 2007 (the latest mortality data at the time of writing) totalled just under 1800. And each year, around 400 000 non-melanoma skin cancers are treated by Australian doctors, costing taxpayers hundreds of millions of dollars.

Many of these patients might have thought their sun exposure was doing them good.

While some sun exposure is vital to good health, Australians in most parts of the country only need a small amount. UV radiation here is harmful to fairer skin types compared with UV levels in most other parts of the world populated predominantly by Europeans. The harms of sun exposure in Australia far outweigh the risks of vitamin D deficiency.

During summer, most of us get adequate vitamin D from just a few minutes' daily exposure to sunlight on our face, arms and hands, on either side of the peak UV periods – before 10 am and after 3 pm.

In winter in the southern parts of Australia, where UV radiation is less intense, people need about two to three hours of sunlight spread over a week. In winter in northern parts of the country, you can still maintain adequate vitamin D levels by going about your day-to-day activities, so there's no need to deliberately seek UV exposure.

Some groups are at higher risk of vitamin D deficiency, such as naturally dark-skinned people, those who cover their skin for cultural reasons, osteoporosis patients, people who are housebound and babies and infants of vitamin D deficient mothers. People in these groups should talk to their doctor about whether they need a vitamin D supplement.

The point is, we can all get our vitamin D without the risk of sunburn.

Does a suntan protect you from sunburn? In most cases, no. A tan can offer very limited protection, but no more than SPF4 (the lowest sunscreen rating), depending on your skin type. A tan does not protect from DNA damage, which can lead to skin cancer.

What about the idea that a tan looks healthy? Although there is a cultural association between tanning and outdoor activities, the reality is that in most cases a tan is a mark of damaged skin. It may not be obvious at first, but over time tanned skin becomes more visibly wrinkled and in many cases patchy and discoloured, compared with skin that has been protected from harmful UV radiation.

People of northern European descent have skin that has not evolved to be suitable for exposure to the Australian sun's intense UV radiation, so a tan is neither natural nor healthy.

Then there's the myth that UV damage to skin occurs predominantly in childhood. Although babies and young children have more sensitive skin than adults, UV damage to your skin at any age increases your risk of skin cancer.

One of the world's most comprehensive studies of sun protection among adults monitored 1600 people in sunny Nambour, Queensland, with an average age of 49 and found that those who regularly used sunscreen over four-and-a-half years developed significantly fewer squamous cell (non-melanoma) carcinomas. Over 10 years, the group applying sunscreen also developed only half as many melanomas as the control group.

So sun protection can reduce your risk of skin cancer at any age.

That's why it's important to slip (on a shirt), slop (on some sunscreen), slap (on a hat), seek (shade) and slide (on your sunglasses), knowing you'll be reducing your skin cancer risk while in most cases still getting enough incidental sunlight for good health.

Fish oil is good for heart health

Michael Vagg, *Clinical Senior Lecturer, Deakin University School of Medicine; and Pain Specialist, Barwon Health*

Did you hold your nose and take your daily dose of fish oil this morning? Or perhaps you opted for an odour-free capsule? Well, you're not alone. Around one in four Australians take fish oil supplements to improve their health. After all, it's supposed to be good for the heart. Right?

The Heart Foundation even has a Fish Oil Program aimed at increasing awareness of the benefits of marine-derived omega-3 fatty acids on heart health. Its 2008 Position Statement on Fish Oil recommended adults consume at least 500 mg of omega-3 a day to lower the risk of heart disease. You can find it at http://www. heartfoundation.org.au/SiteCollectionDocuments/Fish-position-statement.pdf.

If you're interested in learning more about the complexities of fatty acid metabolism, you can find a comprehensive explanation in Wikipedia's entry on omega-3 fatty acid. But in short, researchers are interested in the potential benefits of docosahexaenoic acid (DHA), eicosapentaenoic acid (EPA) and docosapentaenoic acid (DPA).

The fish oil story reflects the challenges involved in translating research evidence into community knowledge and behaviour. And it shows why those who stand by evidence in medicine must be prepared to give up their most cherished beliefs if the science demands it.

The Heart Foundation 2008 Position Statement is a good place to start. It is a serious public health document which takes a balanced, evidence-based approach.

A table in the document sets out the levels of evidence supporting the rationale for its recommendations. But only one finding has level one (the highest level) evidence to support it – that is, the statement that fish oil supplements have a favourable effect on serum triglyceride levels and HDL (good) cholesterol levels.

Although one might assume this to be a good thing, there is not a similar level of support for the direct link between fish oil and improved heart health. It may be that this positive effect on blood lipid ratios is too small to have a useful benefit when applied routinely to real, paying customers outside of clinical trials.

Most of the other planks of the position have level two or three evidence, which translates to positive results in individual studies without support from meta-analysis or systematic reviews of multiple studies. The number of studies with negative or inconclusive results does not affect these ratings.

The Heart Foundation noted this shortfall and designated areas for further study which include doing higher-level reviews and better studies with robust methodology. As it's an organisation which strives to use the best evidence to form policy, I expect it will update and amend its recommendations to reflect the findings of such new studies as they come in.

The biggest gap in the research on fish oil supplements has been between their effects in test tubes and small pilot studies, and the real world of clinical practice. A definite mechanism for how omega-3 fatty acids provide cardiac protection has never been agreed on, which makes predicting clinical effects in real patients difficult.

An early systematic review in 2006 supported fish oil for heart health. But on reading the body of the paper, it's hard to see how the researchers came to such a positive conclusion. They comment throughout on the variable methods of studies reviewed and the mixed outcomes reported. They mostly favour a small positive effect but don't consistently point the same way.

There are also other, similar reviews from the early 2000s which draw mixed conclusions.

More recently there have been major systematic reviews in the *Archives of Internal Medicine*, the *Annals of Internal Medicine* and the *Journal of the American Medical Association* (JAMA) all of which failed to support a clear effect on people in the real world.

The JAMA study demonstrates pretty comprehensively that omega-3 supplements aren't effective at preventing cardiac problems such as heart attacks, stroke, sudden death and arrhythmias. What makes this study more credible is that it included both dietary and supplement studies. Whether you are getting your omega-3s from a capsule or from tins of tuna, it seems unlikely they are doing much good.

So what does this all mean for those taking omega-3 supplements?

Well, fish oil may be reasonable as an add-on therapy for very high-risk cardiac patients who can't tolerate other, more effective treatments.

Research on fatty acid supplementation is likely to continue and some specific use may still be found for such supplements. The evidence that omega-3 can reduce the symptoms of inflammatory arthritis, for example, still appears promising.

Efforts to encourage Australians to eat more fish should push on, because preferring fish to red meat is still a worthwhile change for other health reasons.

But it's becoming increasingly clear that having a quarter of the population on fish oil as a preventive supplement is an unjustifiable expense. I will await the self-regulatory response from advertisers of fish oil products to this avalanche of new evidence with interest, as should the Therapeutic Goods Administration.

Fruit juice is healthier than soft drink

Tim Crowe, Associate Professor in Nutrition, Deakin University

We often hear from health experts and well-meaning parents, that soft drink is terribly unhealthy and we should opt for fruit juice instead. But apart from a few additional vitamins and minerals, there isn't much that differentiates fruit juice from soft drink: both beverages will give you the same sugar and calorie hit.

Let me make an important disclaimer here: fruit juice does have a few redeeming health benefits that make it a little better than soft drink. Prune juice can alleviate constipation, cranberry juice may help reduce the risk of urinary tract infections and many juices contain micronutrients such as vitamin C and potassium.

But these nutrients are found in many other foods. And vitamin C and potassium deficiency are hardly public health issues in Australia.

One of the biggest assumptions about fruit juice is it must be healthy because it's full of 'natural sugars'. Fruit juice does contain natural sugar, which is a mix of fructose, sucrose and glucose, but the quantity (and kilojoules) is on par with soft drinks.

The term 'natural' is also misleading, as the sugar (sucrose) in Australian soft drink is just as natural as that found in Australian fruit juice because it comes from sugarcane. Whether juice is extracted from fruit, or sugar is obtained from sugarcane, both are forms of food processing.

And when it comes to your waistline, that sugar has to be used up or it will eventually result in weight gain. Think of that the next time you're lining up for a super-sized freshly squeezed concoction from your favourite juice bar. That one drink may contain six to 10 pieces of fruit and probably has enough kilojoules to meet more than 10 per cent of your daily energy needs.

While science is still unclear in this area, there is evidence to suggest that feelings of fullness (satiety) after a meal are lower when those kilojoules are consumed in liquid form (especially from more clear-type fluids), rather than as solid food.

This could be due to the rapid transit of the liquid through the stomach and intestines, giving less time to stimulate signalling of satiety. This increases the chance of over-consuming energy with the end result of greater weight gain, or a sabotaging of weight loss.

One study conducted by Deakin University researchers found the more fruit juice Australian schoolchildren drank, the more likely they were to be overweight compared with kids who didn't drink fruit juice. A similar link between increased fruit juice consumption and weight gain was seen in a 2006 study of children from low-income families.

When you're drinking fruit instead of eating it, you're missing out on the pulp that's left behind – and that's where all the fibre is. Fibre is an important nutrient for controlling bodyweight and keeping the digestive tract healthy. But most Australians aren't getting anywhere near the 30 g for men and 25 g for women of fibre recommended by the National Health and Medical Research Council.

Fibre also helps protect against colorectal cancer, the second biggest cancer killer of Australians each year, after lung cancer. In a 2010 update to the most comprehensive report ever published on the role of food, nutrition and physical activity on cancer, the World Cancer Research Fund upgraded the level of evidence linking foods containing fibre with protection against colorectal cancer from 'probable' to 'convincing'.

For someone struggling to keep their weight in check, drinking too much fruit juice or soft drink will make it hard to lose weight or maintain a healthy weight. If you feel the need for a drink, water is your best choice. And when it comes to fruit, eat it, don't drink it.

Hospitals get busier on
full moons

Michael Vagg, *Clinical Senior Lecturer, Deakin University School of Medicine; and Pain Specialist, Barwon Health*

It's another busy night in an inner city hospital emergency department (ED) and patients keep pouring in with injuries from accidents, assaults and self-harm attempts. One veteran nurse turns to a junior doctor, rolls her eyes and mutters, 'Must be a full moon tonight.'

The junior doctor racks her sleep-deprived brain to remember what the moon looked like on the way to work. She can't recall – and doesn't bother checking – but files the incident away so she can bring it up at social gatherings and retell it as part of the folklore of the ED.

What we have just witnessed (or something like it) happens every week around the country. Despite the 'lunar effect' being one of the most easily studied and thoroughly debunked myths in medicine, it provides a fascinating look at the psychology of rational people who work in jobs where superstitious beliefs can easily develop.

But first, the science.

The biggest study I can find debunking the myth is a 2004 paper from Iran (Zargar *et al.* 2004). It analysed just under 55 000 trauma-related admissions to three EDs over 13 months. There was no difference between full-moon days and other days in the number of attendances, nor in the type or severity of patients treated.

Other well-conducted studies include a 1992 Canadian analysis of the relationship between crisis calls to police stations and poison centres, and the lunar cycle (there was no correlation). And another review, also from 1992, also found no relationship between full moons and increased suicide attempts.

A less technical summary from *Scientific American* explains the ancient origins of the belief: the moon was thought to cause intermittent insanity. That's why the Latin word for moon, *luna*, forms the base of the out-dated term lunatic.

Perhaps the more interesting aspect of the *Scientific American* summary is that the myth persists within groups of professionals who deal with unpredictable patients, such as mental health professionals and emergency service personnel.

Two important effects are at work here. The first is confirmation bias. This is the innate tendency of people to remember and pay attention to facts that confirm an already-held belief, and ignore or downplay facts that tend not to support it.

There are many, many busy and stressful nights in ED, and some of them inevitably fall on nights when the moon is full. These will be the ones ED staff remember, and they will tend to forget the ones which are not on full moon nights.

The second likely cause of such a persistent myth is that those who work in fields that are inherently unpredictable – where stakes are high and conditions demanding – are more likely to be prone to superstitious or magical thinking. It is a form of the illusion of control cognitive bias.

Part of feeling able to cope with the randomness of life is to develop associations which can be given a meaning, even if the belief seems absurd. Psychologist Michael Shermer has coined the term 'patternicity' to describe this tendency in the context of evolutionary psychology.

So despite a pedigree dating back to Aristotle, the belief that the full moon affects behaviour in any way only persists because of the very human responses of our front-line health personnel. When they're feeling under the pump, they begin to instinctively look for patterns in random events.

Reference

Zargar M, Khaji A, Kaviani A, Karbakhsh M, Yunesian M, Abdollahi M (2004) The full moon and admission to emergency rooms. *Indian Journal of Medical Sciences* **58**(5), 191–195.

Cracking your knuckles causes arthritis

Michael Vagg, *Clinical Senior Lecturer, Deakin University School of Medicine; and Pain Specialist, Barwon Health*

For some it's a morning ritual – cracking your knuckles before beginning the day. For others, it's a way to pass time while pondering a thought or reading something particularly interesting online.

But are generations of well-meaning parents right? Will knuckle-cracking give us arthritis? Or is it just another harmless habit?

Thanks to Dr Donald Unger MD, this medical chestnut has a definitive answer. In 1998, Dr Unger wrote a Letter to the Editors of *Arthritis and Rheumatism*, the world's premier rheumatology journal.

Dr Unger reported that he had been cracking the knuckles on his left hand at least twice daily over a 50-year period, while the right hand was never cracked and used as a control.

This heroic study was undertaken in response to advice from various concerned relatives who warned him against cracking his knuckles in his early life, and continued with dedication (bordering on pathological) for half a century before the final triumphant publication of the result!

Dr Unger's right hand remained arthritis-free throughout his life. And so did his left.

Fittingly, Dr Unger was awarded the 2009 Ig Nobel prize for Medicine. These awards are presented annually on the eve of the real Nobel Prizes by the organisation Improbable Research for 'achievements that first make people laugh, and then make them think'.

Another earlier study by Swezey and Swezey was published in an obscure journal in 1975 (Swezey and Swezey 1975). The authors were a doctor and his 12-year-old son.

After seeing Unger's letter, they reported their 10-year follow-up had the same result: no arthritis in cracked knuckles.

So why do some of us feel the need to crack our knuckles?

Cracking of a joint is most likely due to a rapid change in joint volume causing the brief formation of a bubble of gases such as carbon dioxide (CO_2).

These gases are normally dissolved in the joint fluid and escape the solution when the pressure of the joint suddenly lowers.

The unstable gas bubble rapidly implodes and is believed to be the cause of the cracking sound.

Joint cracking shouldn't be confused with the snapping sound made by stiff tendons or other bands of soft tissue sliding between muscles or over bony outcrops.

It's also different from the grinding sound (called 'crepitus' by health professionals) that results from movement of a joint with roughened or worn cartilage.

If you've been worried about your knuckle habit, you can relax and get cracking, because the evidence suggests you're not doing any harm to them.

But if you're a chronic workplace or social knuckle-cracker, the harm may come from those who have to put up with you.

Reference
Swezey R, Swezey S (1975) The consequences of habitual knuckle cracking. *Western Journal of Medicine* **122**, 377–379.

Leave leftovers to cool before refrigerating

Clare Collins, *Professor in Nutrition and Dietetics,*
University of Newcastle

Food poisoning doesn't just come from dodgy kebabs, under-cooked chicken and restaurants with poor hygiene practices – it can also occur in the home. And anyone who has suffered a bout of food poisoning knows it's not pretty.

The specific symptoms, and the time it takes until you get sick, vary depending on the pathogen and include nausea, stomach cramps, fever, vomiting and diarrhoea. People who have compromised immune function are particularly susceptible to food-borne illness, including babies, young children, pregnant women and the elderly.

Australia's surveillance system for food-borne illness is monitored by OzFoodNet. You can find them at http://www.ozfoodnet.gov.au/. In 2009 alone, more than 2600 Australians became ill from food poisoning. Of those, 342 required hospitalisation and eight died. OzFoodNet reported restaurants were the most common setting for food contamination.

But many mild cases of food poisoning from home-prepared foods never get reported.

Temperature danger zone

Foods that are cooked then reheated are more likely to be a risk for food poisoning. The greatest potential hazards are meats, casseroles, curries, lasagna, pizza, sauces, custards, patties, pasta, rice, beans, nuts and foods containing eggs, such as quiche.

As cooked food drops to 60°C or below, bacteria that have survived the cooking will start to multiply until the food cools down to 5°C. The longer the food is left to cool, the longer the bacteria – which cause food poisoning – have to multiply.

Food Standards Australia New Zealand (FSANZ) provides a guide to managing potentially hazardous foods in the risky temperature zone: food should take no more than two hours to cool from 60°C to 21°C, and no more than four hours to cool from 21°C to 5°C. You can download the guide at http://www.food-standards.gov.au/_srcfiles/Appendix.pdf. If you want to check this at home, invest in a good quality food probe thermometer (and follow the manufacturer's instructions).

Try to eat food promptly once it's cooked. Or, if you intend to store cooked foods to eat later, you can cool them on a bench as long as the temperature doesn't drop below 60°C. This is roughly when the steam stops rising. You can keep cooked meals safely in the fridge for a few days, but if you want to keep them for longer, put them straight into the freezer.

In the fridge, make sure you store cooked foods on the top shelves and raw foods on the bottom shelves to avoid any contamination from condensation on the raw food that falls onto cooked food.

And finally, when defrosting food, put it in the fridge and keep it below 5°C. Never leave it to defrost on a bench at room temperature because this places it right into the food hazard temperature zone.

When it comes to food safety, a little common sense goes a long way. Always wash your hands before handling food and use separate utensils and chopping boards for raw and cooked food. If you're in doubt about the risk of something you find lurking in your fridge or freezer, throw it out.

Low-fat diets are better for weight loss

Clare Collins, *Professor in Nutrition and Dietetics, University of Newcastle*

If food is labelled low-fat, it's got to be better for weight loss, right? Wrong – it's the total kilojoules that matter most for weight loss. Looking solely at fat content only gives you part of the picture.

Back in the 1970s, few foods were specifically manufactured to be low in fat. If, on doctor's orders, you had to follow a low-fat diet, a trip around the supermarket was fast – skim milk, lots of vegetables and fruit, cottage cheese, fresh fish, and that was about it. Needless to say, you lost weight and got bored very quickly.

In 1982, we got our first set of Australian Dietary Guidelines, recommending we 'avoid eating too much fat'. After that, a few more low-fat products appeared on the supermarket shelves.

The second edition of the guidelines in 1992 called for us to 'eat a diet low in fat, and in particular, low in saturated fat'. This was attributed to recognition that obesity had now become a problem in Australia.

Then, in the mid-1990s, guidelines for foods with low-fat and reduced-fat content claims on their labels appeared in the Code of Practice on Nutrient Claims. From this time we really started to see an increase in products making these claims.

If a food label claims a product is low in fat, it must contain only three grams of fat or less per 100 g of product. If it says reduced fat, it must contain at least 25 per cent less fat than the usual version of this food.

The dietary guidelines were revised again in 2003 and the recommendations for lower total fat and saturated fat choices were embedded across several of the guidelines, with a specific statement to 'take care to limit saturated fat and moderate total fat intake'.

We have now been advised to reduce our fat intakes for over 30 years, and low-fat products flood our supermarkets and food courts. Sadly, however, our waistlines have continued to expand and our fat intakes have not reduced.

Something is clearly awry. Somewhere along the way, low-fat has been misinterpreted to mean 'eat more' because it is low in fat. So rather than linger over a sliver of sponge cake that would contain about 600 kilojoules (150 calories), I can scoff a muffin the size of a small plate with as many as 2000 kilojoules (500 calories), because the sign said low-fat.

Just because a food is low in fat doesn't mean it's low in total kilojoules.

What does the evidence say?

Food is made from macronutrients: protein, fats and carbohydrates. These each provide energy in the form of kilojoules. One gram of protein provides 17 kJ, one gram of carbohydrate provides 16 kJ and one gram of fat provides 37 kJ. Fat has the most kilojoules per gram so you would think that following a diet low in fat should automatically mean you eat fewer kilojoules and lose weight. Not so.

My colleagues and I recently updated the adult weight-management guidelines for the Dietitians Association of Australia and reviewed the recent evidence comparing higher carbohydrate/lower fat diets for weight loss to higher fat/lower carbohydrate diets.

The good news was we found there were lots of studies, with a systematic review and seven recent randomised controlled trials. Overall, when protein and energy intake were held constant across the two diets, both approaches were equally effective in achieving weight loss.

The latest draft of the Australian Dietary Guidelines says to 'limit intake of foods and drinks containing saturated and trans fats and to include small amounts of foods that contain unsaturated fats'.

If you're putting low-fat foods into your shopping trolley, it's wise to stick to the ones that would have been around in your grandparents' day. This means eating more vegetables, fruits and low-fat dairy products, plus whole grains, fish or vegetarian sources of protein such as baked beans. It also means avoiding packaged and highly processed low-fat foods.

If a low-fat diet is not palatable to you, you can still lose weight eating more fat, but you will need to be more aware of the kilojoule value of the foods you choose to eat.

The bottom line is that when it comes to weight loss, watch the total kilojoules, because it's excess kilojoules rather than dietary fat that leads to weight gain.

Men also go through menopause

Carolyn Allan, Endocrinologist, Prince Henry's Institute and Andrology Australia

Feeling tired and grumpy? Maybe a little emotional? If you're a middle-aged male, these symptoms might be hormone-related, but no, you're not going through *man* opause.

It's true that signs of men's low testosterone are similar to symptoms of menopause: low energy, mood swings, irritability, poor concentration, reduced muscle strength and bone density, and a lack of interest in sex.

But unlike women, whose oestrogen levels fall rapidly when they go through menopause, men's testosterone declines much more modestly and gradually.

Testosterone levels in men are highest between the ages of 20 and 30 years, and from 30 to 80 years they drop by around a third. Some men will experience a greater drop than others, particularly if they're overweight or obese.

Testosterone is essential for good health because it stimulates the growth of muscles, bones, and the bone marrow that makes red blood cells.

So testosterone or 'androgen' deficiency – which affects one in 200 Australian men under 60 – can have a major effect on the body's ability to function normally.

The 'manopause' myth has been perpetuated by the interest in testosterone replacement therapy (TRT) as an elixir of youth to improve the symptoms and signs of ageing.

Testosterone therapy offers benefits for men with known causes of androgen deficiency, but there is a lack of data to define the level of deficiency that warrants this treatment.

If the cause of androgen deficiency is unknown, treatment needs to be tailored to the individual. Testosterone treatment certainly shouldn't be requested or prescribed in the belief that it's a 'cure-all' for symptoms of ageing.

For ageing men without classic androgen deficiency the jury is still out on the effectiveness of testosterone replacement therapy.

The safety of the treatment – particularly in the case of the prostate and cardiovascular system – is unclear, and the benefits seem relatively modest. There is certainly no remarkable return to youthful vigour.

Often, low testosterone levels can be a sign of underlying health conditions. Low testosterone levels are associated with various chronic diseases such as diabetes, heart disease and depression.

The Massachusetts Male Ageing Study (the largest study of male ageing) found that chronic illness, use of prescription medication, obesity or excessive alcohol intake were associated with a 10 to 15 per cent reduction in serum testosterone levels (testosterone in the blood) in men aged over 40 years.

Treatment, then, should focus on reducing the risk factors for these conditions (i.e. weight loss, reduced alcohol intake) rather than the testosterone level.

While the idea that men, too, go through menopause might be a playful explanation of the ageing process, it shouldn't be taken too seriously, especially if serious symptoms of chronic diseases are dismissed.

The MMR vaccine causes autism

Rachael Dunlop, Postdoctoral fellow, University of Technology, Sydney

Few medical myths have spread as feverishly and contributed to so much preventable illness as the theory that the triple measles, mumps and rubella (MMR) vaccine might be linked to autism.

The tale was first suggested by Andrew Wakefield at a 1998 press conference following the publication of his now discredited (and retracted) paper in *The Lancet*.

The paper itself didn't address such a connection but Wakefield raised concerns with journalists and called for a boycott of the MMR vaccine.

'I can't support the continued use of these three vaccines, given in combination,' he said, 'until this issue has been resolved.'

Wakefield said the vaccine should instead be broken into single components and given at yearly intervals.

We now know Wakefield had good reason to discredit the MMR: he had a patent for a single measles vaccine and he was being paid by lawyers who were assembling a case against MMR manufacturers.

None of these conflicts of interest was revealed when the *Lancet* paper was submitted for publication – if they had been, it would never have been published. As the editor of *The Lancet* noted, Wakefield's paper was 'fatally flawed'.

Further investigation published in the *British Medical Journal* revealed that what Wakefield did wasn't just bad science, but deliberate fraud.

Wakefield was struck off the UK medical register in 2010 for 'callous, unethical and dishonest' behaviour. But the damage had already been done.

A drop in MMR vaccination rates lead to inevitable outbreaks of preventable disease, and measles is again endemic in Europe.

The episode also prompted research on possible links between MMR – and vaccines in general – and autism. Now, more than a decade after Wakefield's paper was published, we have considerable evidence that MMR is not linked to autism.

One of the largest single studies to look for a link came from Denmark and covered all children born from January 1991 to December 1998. The study examined a total of 537 303 children, 82 per cent of whom had been vaccinated for MMR.

It found no association between vaccination and the development of an autistic disorder.

More evidence comes from Japan, which stopped using the trivalent vaccine in 1993 over safety concerns with the anti-mumps component of the MMR formulation.

A study of more than 30 000 children found autism cases continued to rise even after the MMR was withdrawn and replaced with single vaccines, providing strong evidence that the MMR vaccine was not implicated.

More recently, the United States Institute of Medicine completed an exhaustive review of all the scientific literature and concluded there was no causal relationship between MMR vaccine and autism.

So science has rejected such a link, but what have the courts found?

The US Court of Federal Claims (Vaccine Court) was established in 1988 as a no-fault system for litigating vaccine claims.

In 2007 the court began to hear the 'autism omnibus' trials – a class action of almost 5000 lawsuits attempting to demonstrate MMR played a causal role in the development of autism.

The group put forward the best three cases as a trial and the decision was handed down in 2010.

Judge Hasting wrote of one case, 'Considering all of the evidence, I found that the petitioners have failed to demonstrate that … the MMR vaccine can contribute to causing either autism or gastrointestinal dysfunction.'

Patricia Campbell-Smith, special master on another case, said 'The petitioners' theory of vaccine-related causation is scientifically unsupportable.'

This myth has been well and truly busted.

Science still doesn't know exactly what causes autism, but researchers are continuing to look.

In the meantime, it's important that parents get accurate information about vaccines so they can protect their kids from preventable disease and avoid getting taken in by expensive and dangerous quack therapies such as chelation therapy, which aims to remove heavy metals from the body.

Who knows, if we hadn't been sent on a wild goose chase by the nefarious research of Andrew Wakefield we might be closer to understanding this syndrome.

MSG is a dangerous toxin

Merlin Thomas, *Adjunct Professor, Preventive Medicine, Baker IDI Heart and Diabetes Institute*

Monosodium glutamate (MSG) is widely viewed as a dangerous food toxin that is responsible for adverse reactions to Chinese food and other meals. But is it really the MSG that's to blame?

Glutamate is a naturally occurring amino acid, used chiefly in our body to make protein. Most of us normally eat around 10 g of glutamate every day, much of which is released into the body when we eat and digest protein. Some protein-rich foods – such as meat, fish, chicken, dairy products, legumes and corn – are especially rich in glutamate.

But glutamate has another important property: it tastes good. When glutamate touches the taste receptors on our tongue, it gives food a savoury taste (known as umami). Mixed with our meal, glutamate is said to balance, blend and enhance the total perception of flavour. But not just any glutamate; we can't taste glutamate that is locked in protein. For it to tickle our taste buds, it must be in a 'free form'.

Some (tasty) foods – tomato products, fermented soy/fish/oyster/steak/Worcestershire sauces and long-matured cheeses such as stilton and parmesan – are high in free glutamate. Ever wonder why parmesan makes the bolognese taste better (and more meaty)? It's the glutamate!

Because MSG has such a bad name, many manufacturers use other sources of glutamate to give processed foods the extra taste. These include vegetable, corn, yeast or soy protein extracts, in which the glutamate has been released from the protein by enzymatic digestion or chemical hydrolysis (a reaction which uses water to break the bonds of substances). When dissolved in water, the free glutamate in these extracts is chemically identical to that contained in MSG and enhances flavour in precisely the same way.

Most of us would usually eat around half to one gram of free glutamate every day as additives to our food. In Asian countries, this figure is double, reflecting the use of soy and other fermented products in cooking. A highly seasoned banquet in a Chinese restaurant may contain up to four to five grams of free glutamate.

But glutamate isn't just found in Chinese restaurants. Many American-style fast foods contain just as much glutamate to enhance their flavour (and your experience) beyond that of their competitors. Even Vegemite contains 1.4 per cent free glutamate.

A small proportion of people experience transient symptoms when they consume large amounts of free glutamate (more than four to five grams) in a single meal. These reactions vary from person to person but may include headaches, numbness/tingling, flushing, muscle tightness and general weakness.

Several scientific studies have tried to replicate this experience. Most have been too small, used unrealistically high doses of MSG, and were not undertaken in the context of food (or even with intravenous doses). Try eating a whole jar of Vegemite in one sitting and you will soon see why people don't feel so well afterwards.

Other, more rigorous studies have failed to confirm a reproducible response to meals containing MSG, even in self-attributed 'MSG sensitive' individuals. Most reactions to a Chinese banquet probably have little to do with the MSG, as many of the same people who are 'MSG sensitive' have no problems with Vegemite or parmesan cheese.

It has also been suggested that MSG can trigger an asthma attack. While there are lots of anecdotal reports (again, usually after Chinese food), challenge studies with MSG-rich meals have generally failed to confirm these findings. There are many other things in food that can trigger an attack in sensitive individuals, from dairy products, eggs, peanuts and sulphites, to food colourings. But none is vilified like MSG.

Finally, it has also been suggested that MSG leads to weight gain and obesity. Of course we have a great tendency to eat more of anything that tastes better, so this comes as little surprise. MSG has even been used to promote the appetite of frail elderly people.

The consensus among clinicians and scientists is that MSG is safe for human health. Very high doses may affect some people for a short time but there may be far more dangerous consequences that come from overeating.

Natural cancer therapies can't harm you

Ray Lowenthal, Professor of Oncology, University of Tasmania

One of the most misleading myths of modern medicine is that conventional cancer doctors reject 'natural' therapies in favour of artificial or 'unnatural' cancer treatments. This myth has contributed to the popularity of unproven, alternative cancer treatments.

The truth is that oncologists and other trained medical professionals involved in cancer care welcome and support effective cancer treatments in any form, provided there is evidence to show they can work and are safe.

Making assumptions about the benefits and harms of therapies according to whether or not they are natural is high-risk. For example, laetrile, an extract from apricot kernels, was for years promoted as a natural alternative therapy for cancer; yet it is utterly useless for treating cancer and can cause fatal cyanide poisoning.

The herb comfrey, also recommended as an alternative cancer treatment, can actually *cause* cancer.

So 'natural' does not necessarily equate to 'harmless'. Nor does conventional necessarily equate to unnatural. Plenty of natural products are used in chemotherapy. These include extracts from the yew tree (docetaxel, paclitaxel), the opium and mandrake plants (epipodophyllotoxins) and from natural moulds that produce doxorubicin and related drugs, used effectively to treat breast cancer and lymphoma.

Some natural products used in conventional cancer medicine had for centuries been part of traditional folk remedies and have been adapted for modern use after being rigorously tested.

So the difference between alternative and conventional is not that one is natural and the other is not. It's that conventional cancer treatments must be subjected to rigorous research before they can be recommended for use and prescribed by professional oncologists.

The highest level of research is the randomised control trial, which is only applied to a product after lengthy laboratory studies, preliminary testing and approval by an ethics committee made up of medical experts, ethicists and healthcare consumers.

A typical trial involves randomly selecting two groups of patients in large enough numbers to control for physical differences between them. One group receives the new treatment and the other group is given a different treatment or a placebo; the results are then compared. A trial is designed to show that any significant difference in patient outcomes can only be the result of the treatment being tested.

Oncologists will only prescribe treatments if they have been tested in this way and are found to be effective and safe.

A good example of this testing process on a natural derivative is the development of the drugs vincristine and vinblastine, extracted from the Madagascan periwinkle. Improved through continual clinical trials over 50 years, these so-called 'vinca' alkaloids have been a key part of modern-day successes in curing childhood leukaemia and other cancers that were previously incurable.

Some alternative cancer therapists also promote fad diets, but there is no evidence to support this approach. A healthy diet can prevent cancer and assist people living with cancer. But diet will not cure cancer, which directly attacks the body's cells in a highly destructive and relentless way.

Such a malignant disease can only be cured if the cancer cells are surgically removed before the cancer has spread or if they are destroyed using chemotherapy and/or radiotherapy.

Nor is there any evidence to support mind control in any form as a cancer therapy. Such a belief or expectation in many cases adds to a patient's distress. Can you imagine the terrible trauma of being diagnosed with a potentially fatal cancer and told you can think your way to good health with a positive attitude?

The reality is, we have a limited life span; science does not have all the answers to our health needs. But we agree as a society that we should do what we can to increase life expectancy and improve health.

Over the past century, average Australian life expectancy has increased by almost 30 years, largely through a combination of improved infection control, sanitation, diet, immunology and many other advances in medical science.

The changes in medical practice and public policy that have improved our length and quality of life were guided by evidence of what works.

So we must let the evidence – not uninformed perceptions of what is natural – guide continuous improvements in cancer treatment.

No pain, no gain

Peter Milburn, *Professor, School of Rehabilitation Sciences, Griffith University*

The value of regular physical activity to a person's well-being is unequivocal. But how much exercise do we need to maintain health, improve fitness or lose weight? And where is the line between healthy and harmful?

Australia's dietary guidelines recommend adults do at least 30 minutes of moderately intense physical activity – such as brisk walking, social tennis or swimming – on most days to maintain a healthy weight. But if we want to lose weight, and don't cut back on food and drink, we need to do more.

The American College of Sports Medicine agrees adults should get at least 150 minutes of exercise a week, though it explains this might be 20 to 60 minutes of vigorous exercise – which makes you huff and puff, such as jogging, aerobics, football and netball – three days a week.

The College guidelines also prescribe the quantity and quality of exercise for developing and maintaining cardiorespiratory, musculoskeletal and neuromotor (or functional) fitness in healthy adults:

- For flexibility, adults should do stretching exercises at least two days a week, with each stretch being held for 10 to 30 seconds, to the point of tightness or slight discomfort.
- For resistance training, adults should train each major muscle group two or three days a week.
- For cardiovascular fitness, people should gradually increase the time, frequency and intensity of their workout.

The belief that 'if a little bit of exercise is good for me, then more should be better', still pervades the fitness industry. As does the 'no pain, no gain' myth, which came to prominence in the early 1980s via Jane Fonda aerobic workout videos. Fonda would also urge viewers to 'feel the burn' and exercise beyond the point of reasonable physical stress. These days the 'no pain, no gain' motto is used to show that physical development is the result of training hard.

We often judge the efficacy of our workouts by our level of soreness the next day. This type of pain is called delayed onset muscle soreness (DOMS) and occurs a day or two after exercise. It is most frequently felt when you begin a new exercise program, change your routine, or dramatically increase the duration or intensity of your workout. DOMS is a normal response to unusual exertion and is part of the body's adaptation process that leads to increased strength or endurance as muscles recover and hypertrophy.

But while discomfort is natural if you push yourself, pain is the body's protective mechanism, warning us to ease the intensity or protect an injured part of the body. Resisting this warning risks damaging tissue and may cause your body to over-compensate with other movements that can aggravate the injury and lengthen healing time. It's also likely to reduce your motivation to continue exercising.

Pain during exercise can also indicate underlying health problems and should be seen as a signal to stop exercising and seek professional advice:

- Chest pain during exercise is a red flag for potential heart problems.
- Exercise-induced bronchospasm (a sudden constriction of the bronchial muscles), even in non-asthmatics, may indicate an underlying respiratory problem.
- Joint pain may result from osteoarthritis or indicate meniscal (knee) injury, or ligament or tendon microdamage.

If you do find yourself sore after a tough workout or competition, try some low-impact aerobic exercises to maintain your blood flow during warm-down. Other remedies such as massage, ice baths and the RICE (rest, ice, compression, elevation) combination may also ease muscle soreness.

Medication such as non-steroidal anti-inflammatory drugs like aspirin and ibuprofen can temporarily help reduce the effects of muscle soreness, though they won't speed up healing.

It's certainly not easy building up your fitness or losing weight but the 'no pain, no gain' motto is based less on the science of exercise physiology than on outdated sports psychology. It's a recipe for injury.

When you feel pain during exercise, stop what you're doing and take stock of how you're feeling. If you think you can, try returning to the activity you were doing, but if the pain persists, then stop for good.

Organic food is more nutritious

Clare Collins, Professor in Nutrition and Dietetics,
University of Newcastle

Across the world, outbreaks of food-borne illness, contamination and environmental scares have generated a lot of media attention and plenty of fear around food safety. Think of *E. coli* outbreaks, the Fukushima radiation leaks, health concerns about pollutants and the use of pesticides.

The resulting distrust of the food supply has driven consumers to seek what they believe are the safest, cleanest, most sustainable and nutritious food choices: organics. And this is the type of food you would want to feed those you love, your family, your baby and yourself.

The truth is that neither organic nor conventional food is better or worse than the other. But as the demand for organic food increases, so does the myth that it must have higher nutrient values.

Whether you eat organic or non-organic produce is up to you. But factors to consider are cost, nutrition, taste and environmental impact.

Cost

Organic food does cost more than other food – sometimes a lot more. This is due to the extra costs to produce organic food, the need for more labour and, often, smaller crop sizes.

Around 86 per cent of Australians don't eat the recommended five serves of vegetables daily and 46 per cent don't eat the recommended two serves of fruit a day.

Cost can be a major barrier to increasing your fruit and vegetable intake. But it's better to eat twice as much non-organic pumpkin, carrots, zucchinis and beans than half the amount of organic ones.

Eating too few servings of fruit and vegetables is estimated to account for 1.4 per cent of the total burden of disease in Australia, while greater consumption predicts better health and protects against chronic conditions such as high blood pressure, high cholesterol and heart disease.

In Australia, 11 per cent of all cancers are thought to be related to inadequate fruit and vegetable consumption.

Nutrition

In 2009, the prestigious *American Journal of Clinical Nutrition* published a systematic review of the best available evidence evaluating research on the nutritional quality of organic foods.

The researchers searched more than 50 000 articles to locate 162 studies comparing organic and conventional crops and livestock. They found that overall, conventional crops were higher in nitrogen (due to nitrogen fertilisers) and organic crops were higher in phosphorous (due to phosphorous fertilisers) and what they called 'titratable acidity' which is due to ripeness at harvesting.

But there were no differences in other vitamins or minerals analysed, including vitamin C, phenolic compounds, magnesium, potassium, calcium, zinc and copper.

A major shortcoming of the studies, however, was that half had failed to indicate what body certified the crop as organic. The conclusion was that organic and conventional crops were comparable.

Taste

A study published in 2007 took the same varieties of lettuce, spinach, rocket, mustard greens, tomatoes, cucumber and onions. The researchers grew two crops alongside each other in separate plots: one was grown organically and the other using conventional agricultural techniques.

They then asked consumers to evaluate the taste and how much they liked the two types. No differences could be detected, except for a preference for the conventional tomatoes that were slightly riper.

Environmental impact

The environment is the main winner when it comes to organic farming. This is because synthetic pesticides and herbicides are not used, and crops are commonly rotated, so biodiversity is promoted.

The confusion around organics comes from mixing the separate issues of nutrition, taste, safety and environment into the same argument.

Eating enough fruit and vegetables

The best way to improve the nutrients you get from vegetables and fruit is to eat more of them. Go for those that are in season and visit a growers' market to discover those that are grown close to home.

You should aim to eat five serves of vegetables and two serves of fruit each day. One serve of vegetables equals 75 g or half a cup of cooked vegetables, or one medium potato, or one cup of salad vegetables, or half a cup of cooked legumes (dried beans, peas or lentils). One serve of fruit equals 150 g of fresh fruit, or one medium-sized piece, or two smaller pieces (stone fruits, for instance), or one cup of canned or chopped fruit.

Try these suggestions to increase your fruit and vegetable intake.

At breakfast:
- add stewed fruit or a chopped banana to cereal
- grate an apple into rolled oats before cooking
- top toast with cooked mushrooms, tomatoes, capsicum or sweet corn
- heat chopped leftover vegetables and serve as a topping for toast; add an egg or reduced-fat cheese to make a meal.

For snacks:
- cut up a fruit platter
- pack fresh fruit into your lunchbox
- top English muffins or crumpets with tomato paste, diced vegetables and sprinkle with reduced-fat cheese for a quick mini pizza
- serve carrot and celery sticks, florets of broccoli and cauliflower, and strips of capsicum with a low-fat dip

- grate or dice onion, carrot, zucchini, potato and corn into a savoury muffin or into pikelet mixture.

At main meals:

- make meat go further by adding extra vegetables to a stir-fry or a casserole
- add extra vegies, dried peas, beans or lentils to recipes for meatloaves, patties, stuffing, soup, stews and casseroles, pies, nachos, pasta and rice dishes, pizza and pancakes
- serve main meals with cooked vegetables and a salad of baby spinach leaves, cherry tomatoes and olives
- use capsicum, zucchini, pumpkin, eggplant, cabbage and lettuce leaves as edible containers and fill with savoury toppings
- for easy wedges, cut potato, sweet potato, pumpkin and parsnip into wedges; microwave until cooked; mix with a teaspoon of vegetable oil, dried mixed herbs and cajun seasoning and bake in a hot oven until crispy.

Osteoarthritis can be 'cured'

Michael Vagg, *Clinical Senior Lecturer, Deakin University School of Medicine; and Pain Specialist, Barwon Health*

Switch on daytime television on any given day and you'd be forgiven for thinking there was a cure for the debilitating and dreaded condition, osteoarthritis.

Unfortunately, there's not. And that's not from want of trying. With the exception of the common cold, no everyday health problem has been as extensively studied with such little result.

Osteoarthritis is the most common form of arthritis, affecting around 1.6 million Australians. The hallmark of osteoarthritis is loss of cartilage which lines major joints – this causes the classic symptoms of pain and stiffness in the affected joints.

Or does it? One of the major frustrations in osteoarthritis research is the absence of research showing the correlation between the state of joints on X-rays and the degree of pain and disability the operator of the joint experiences.

To work out what creates the pain in osteoarthritis, researchers need to look beyond the joint.

Many studies using magnetic resonance imaging (MRIs) suggest the bone marrow around the joint is a potent source of pain. And further away, in the spinal cord and brain, there may be abnormal processing of pain signals.

When you add the standard physiological, genetic and psychological complexities common to all types of long-term pain, you begin to see why finding durable, meaningful relief from osteoarthritis pain is so difficult.

What osteoarthritis sufferers don't need is to be presented with a steady stream of fake and cynical products which reflect none of this hard-won knowledge.

Tabloid media, particularly television, often feature products promising relief from arthritis pain. Typically the product is presented as an infomercial, with no critical analysis from the reporter. A couple of testimonial cases are then wheeled out for breathless adoration.

There is also invariably some type of pseudoscientific angle presented as an explanation for the miraculous healing powers of the product. It's often cast as a 'secret breakthrough' discovered by a lone misunderstood genius, who is persecuted by vested interests in the pharmaceutical industry.

So how does this type of misinformation make its way onto our television screens and magazines?

The Therapeutic Goods Advertising Code contains what would be robust protection for consumers – if it were adequately policed.

But the Code is held in such low regard by product manufacturers and distributors that some don't bother to follow it. Others don't do their research to understand the Code before tipping thousands of dollars into promoting their questionable devices or pills.

Many of the less scrupulous operators simply ignore TGA sanctions, or make the minimum required changes to their advertising, while maintaining claims about efficacy.

Cynics can chuckle that there's no great harm in condoning this thriving industry built on deceptive advertising. They may even say it's a tax on gullibility.

But it makes things much harder for those of us whose careers involve trying to steer these osteoarthritis sufferers towards the proven interventions that can reduce their disability and pain.

There aren't any miracle cures for osteoarthritis but there are evidence-based measures to treat the condition or reduce its severity. These include:

- peer-led education groups;
- carefully tailored weight loss;
- cognitive-behavioural treatment;
- judicious medication use; and
- joint replacement.

These measures aren't glamorous and challenge sufferers' beliefs about their pain and their lifestyle. This type of change is hard for individuals to contemplate, and difficult to resource and implement.

For the average Australian without much scientific background, it's easy to see the attraction of a simple and compelling story about a miraculous cure.

So what's the solution?

Well, for a start, health-care professions should do more to educate consumers about dodgy arthritis products and avoid lending our professional credibility to endorsing them.

And the government needs to step in with some ruthless regulation.

Peanuts in pregnancy cause allergies

Monique Robinson, Associate Principal Investigator, Telethon Institute for Child Health Research, University of Western Australia

Anyone else have the feeling something radical has happened with peanut allergy in the past 30 years? I don't recall knowing anyone allergic to peanuts or peanut butter as a child in the 1980s, yet today every school is equipped with EpiPens and detailed peanut-response strategies.

And for good reason too. Peanuts are the most common cause of severe allergic reactions to food, including anaphylaxis, the name given to the rapid onset of allergic reactions all through the body including throat swelling, itchy rashes and wheezing. Anaphylaxis will often result in a trip to hospital; at worst it can be fatal, and the swift onset of symptoms makes it extremely frightening for children and parents.

Global data back up this inkling that peanut allergy is increasingly common in the current generation of children. Australian research reports a fivefold increase in hospitalisations for anaphylaxis due to food allergy between 1994 and 2005, particularly in very young children.

So peanut allergy is on the rise, and it's scary. Even worse, we don't yet know what is causing it. It's the perfect storm for the creation of a medical myth that eating peanuts while pregnant causes allergies.

Where did the myth start? In 1996, British researchers observed that infants who had not been given peanuts before showed allergic reactions to peanuts on the first ingestion.

Therefore, perhaps the maternal consumption of peanuts or peanut products during pregnancy was causing the child's allergy.

It was just a suggestion, without any evidence as to what mechanism might lead from mum's peanut ingestion to peanut allergy in the unborn child. However, remarkably, in 1998 the United Kingdom's Committee on Toxicity of Chemicals in Food released advice based on this study stating that pregnant or breastfeeding women who have allergies themselves (or with close relatives with allergies) should avoid eating peanuts and peanut products during pregnancy and lactation.

As you can see, it's a complicated guideline where simple messages are needed. And so it was inevitably shortened, and from 1998 to 2008 while the guideline was in place, around 60 per cent of women, regardless of allergic status, avoided peanuts during pregnancy. First-time mothers were particularly vigilant, being twice as likely as women on second or subsequent pregnancies to avoid peanuts.

Unlike some pregnancy myths, this one is not entirely without basis. There does seem to be some role for maternal transmission of peanut allergy when the mother is allergic to peanuts herself. And while there are one or two studies that support the link between mum's peanut consumption and increased risk for peanut allergy in the baby, there are far more studies that have found this is not the case and many that have found peanut ingestion during pregnancy can reduce the risk of peanut allergy in the child.

The problem is, it's complicated. As anyone with peanut allergy can tell you, it is extremely hard to avoid peanuts entirely, so mothers who reported not eating peanuts may have been exposed to peanuts in other foods or even through body creams. Scientific studies in the area use words such as 'insidious' and 'occult' when describing peanuts, and in my career researching pregnancy risk factors, I can't recall reading such evocative descriptions before. Such is the sneakiness of the peanut.

Then there's the question of how many peanuts does a peanut allergy make? And does it matter when they were eaten – early or late in pregnancy? And most importantly, what about genetic links and other factors in the mother's environment or the child's post-natal world that might have led to the allergy developing?

We just don't know the answers to these questions, but there were enough studies refuting the mother's ingestion of peanuts during pregnancy as the cause of child peanut allergy for the UK guideline to be withdrawn in 2008. Also, time has shown that there was an increase in the diagnosis of peanut allergy while the guideline was in place and women were avoiding peanuts while pregnant and breastfeeding.

So at this stage, there is no evidence to support the claim that eating peanuts or peanut butter during pregnancy will make your child allergic to peanuts. As a mother, you'll have plenty of things to feel guilty about, but being the cause of your child's peanut allergy all because you like peanut butter on your toast at breakfast is not one of them.

Should we stay away from peanuts and peanut products anyway while pregnant just to be on the safe side? Probably not, especially given the research suggesting consumption of peanuts might actually build tolerance in the child.

The best advice might be to eat as you would normally, except the foods that Food Standards Australia New Zealand and your doctor recommend you stay away from.

The placebo effect only works on the gullible

Michael Vagg, *Clinical Senior Lecturer, Deakin University School of Medicine; and Pain Specialist, Barwon Health*

If you took a pill that had been prescribed to treat your illness and it alleviated your symptoms, that means the medicine worked – right?

What if you took a complementary medicine from a health food store instead?

Or if you were given a sugar pill and you still felt better?

Medicines, surgery and alternative therapies all inevitably cause some degree of placebo response – almost invariably in subjectively measured indicators – but it's not just the gullible or suggestible who are affected.

First, you might experience what's known as the Hawthorne Effect, in which people modify their behaviour because it's being studied.

It goes something like this. If you've identified a problem with your health and sought help for it, you will tend to be making other, related decisions which can reduce your symptoms – or your perception of them.

So if you feel your knee becoming sore, you'll probably tend to use it less and be more careful about what you do. Both may help reduce your pain, independently of the treatment you seek.

Having an involved and attentive health professional to empathise with your concerns will likely strengthen your tendency to get on with life instead of fretting about the pain in your knee and what it might mean.

Second, you might succumb to the phenomenon of Regression to the Mean, which is a scientific-sounding way of explaining that any long-term complaint will tend to vary around its average level. Some days it will be a bit worse, other days it will feel a bit better than average.

Our inherent bias is to ignore the good days and focus on the bad. If we string together a few days of worsened symptoms, which will happen at some stage, you might be tempted to try something out of the ordinary – such as alternative therapies – to feel better again.

But the longer the symptom is above or below its long-term average, the more likely it is to begin returning to the usual level.

When this return coincides with the trial of a new treatment you saw on TV or bought over the internet, the treatment seems to work. So you would rationally (but incorrectly) assume you responded to the treatment.

This logic also applies to self-limiting conditions which will last for a set period – whatever treatment is applied – such as back pain or even a cold.

So what proportion of people are likely to experience the placebo effect?

Sham surgery (or placebo surgery) has the biggest placebo response, with some studies reporting positive results in 70 per cent or more of the participants, and long-lasting improvements in function.

This is followed by physical treatments such as acupuncture or Transcutaneous Electrical Nerve Stimulation (which uses electrodes taped to the skin near the site of the pain). Pills are next. Pills can vary from low to high response rates depending on the condition being studied, and active drugs such as morphine can be more effective if delivered openly instead of being hidden.

The size of the placebo effect in psychiatric research appears to be increasing over time, making it harder to trial antidepressant drugs.

This is likely due to more sophisticated study designs which are more sensitive to bias, and to changing community expectations about the effectiveness of the therapies being trialled. It's not likely to be because people are becoming more suggestible.

Deception is supposedly at the heart of the placebo response, but it seems that even this may be a furphy. The strength of patient-health provider interactions can be such that it's possible to get relief from symptoms when knowingly taking a placebo, if patients are primed appropriately with a positive description.

Interestingly, the placebo response doesn't affect everyone. Placebo pain relief is significantly reduced in people with damage to the frontal lobes, such as stroke survivors, brain injury victims or those with dementia.

The placebo response has social, psychological and biological implications that researchers are still struggling to understand. It's certainly not as simple as being fooled into feeling better.

Don't worry, kids will grow out of their 'puppy fat'

Louise Baur, *Professor of Paediatrics, University of Sydney*

Picture this common scenario: a mother is worried about the size of her 13-year-old daughter, who appears quite a bit heavier than the other students in her class. But the mother's friend reassures her that it's only puppy fat and her daughter will grow out of it. So no efforts are made to examine, and potentially alter, the girl's diet or levels of physical activity.

So is 'puppy fat' a true phenomenon? Do young people who are overweight during puberty usually grow out of it?

Let's consider the evidence.

A 2010 study that monitored the weight status of 900 children in Victoria found those who were carrying excess weight in primary school were likely to still be carrying the excess weight in later high school.

The children were first seen between the ages of five and 10, and were followed up eight years later. One in five of the students was persistently overweight or obese in both mid-childhood and adolescence, with some additional young people developing excess weight for the first time in their teen years.

If excess weight in adolescence wasn't associated with any health problems, there wouldn't be cause for concern. But adolescents who are overweight or obese are more likely to have a range of risk factors for heart disease and related problems, just as they are in adulthood.

In a NSW study of almost 500 school students aged 15 years, overweight or obese boys were more likely to have elevated blood pressure and cholesterol, abnormal levels of insulin (suggesting a form of pre-diabetes) and poor liver function, than boys whose weight was in the healthy range.

Similar findings were found among adolescent girls, although they weren't as pronounced.

These health problems, just like excess weight in childhood, don't just go away. Adolescents who are overweight or obese, or who have risk factors for heart disease, tend to retain these health problems into their middle age.

A 2011 Finnish study found teen body mass index (BMI) and disease predictors such as high blood pressure correlated strongly with their risk of obesity-related disease at age 30 to 45.

Given all this, it's clear the popular concept of 'puppy fat' as being just a transient phenomenon is a myth. And Australian children have grown increasingly overweight and obese over the past two to three decades.

What, then, should be our response?

Well, extreme measures such as mandating extra PE classes for overweight students or forcing parents to attend healthy lifestyle programs aren't the answer. But we can't ignore the often life-long health problems obesity brings.

Instead, we need to provide supportive environments for all young people, whatever their weight status. Within the home, we need to encourage parents to provide a healthy food and activity environment.

If a young person does have a weight problem, then this needs to be handled sensitively, with parents supporting teens, rather than forcing them, to make healthier lifestyle choices.

Parents should try not to nag adolescents about what they're eating, as this can put the defences up. But it's a good idea to keep tempting foods such as soft drink, chocolate, biscuits and cakes away from the house. (If they are there, they will be eaten!)

There are some great resources available to assist parents through this process – you can download some excellent tips from The Children's Hospital at Westmead at http://kidshealth.schn. health.nsw.gov.au/fact-sheets/weight-management-tips-parents-helping-your-adolescent-lose-weight. But if you remain concerned about your teen's weight, consider a medical review by a youth-friendly GP.

More broadly, all of us should be advocating for changes to the environment that make healthy choices – around what you eat and how active you are – the easy choices for everyone.

We shouldn't be dismissive of the consequences of excess weight, and especially obesity, in young people. Instead, we need to put to death the myth of puppy fat and address the factors that promote unhealthy weight gain.

Reading from a screen harms your eyes

Harrison Weisinger, *Foundation Director of Optometry and Chair in Optometry, School of Medicine, Deakin University*

The time most of us spend looking at a screen has rapidly increased over the past decade. If we're not at work on the computer, we're likely to stay tuned into the online sphere via a smart phone or tablet. Shelves of books are being replaced by a single e-book reader; and television shows and movies are available anywhere, any time.

So what does all this extra screen time mean for our eyes?

Well, you'll be pleased to hear that as with many good eye myths, there is simply no evidence to support this old chestnut.

Once we reach the age of 10 years or so, it is practically impossible to injure the eyes by looking at something – the exception, of course, being staring at the sun or similarly bright objects. Earlier in life, what we look at – or rather, how clearly we see – can affect our vision because the neural pathways between the eye and brain are still developing.

When we read off a piece of paper, light from the ambient environment is reflected off the surface of the paper and into our eyes. The retina at the back of the eye captures the light and begins the process of converting it into a signal that the brain understands.

The process of reading from screens is similar, except that the light is emitted directly by the screen, rather than being reflected.

Some people worry about the 'radiation' coming from screens but there's nothing unhealthy about it. The radiation is, for the most part, just visible light, which is why we can see the screen in the first place. Most of the other emissions that lie outside of the visual spectrum are either low energy and unharmful, or absorbed by one of the front few layers of the eye, including the tear film.

In the past, people used to buy screen covers to dim the light being emitted from their screen. I suspect this did little more than dim the light – causing them to squint and strain. I like bright, shiny screens but the choice between shiny and matte screens is really only one of personal preference.

Many people complain that prolonged periods looking at a screen give them headaches and sore eyes. This is perhaps a reflection of the fact that, when looking at a screen and focusing on nearby objects, our eyes are not really doing what they've been designed for. The eye evolved predominantly to be able to look out over fields for potential food or for hungry lions, with the occasional requirement to look at things up close.

We can look up close when the lens inside the eye 'accommodates'. This requires contraction of muscles inside the eye. When we fixate on a nearby object (say, a screen), we also must turn our eyes inwards. This is called convergence.

With hours on a screen, the muscles of accommodation and convergence can fatigue and give rise to the symptoms we know as eye strain. In my experience, this is one of the most common causes of headache in people who work on screens all day.

This is not to say that screens cause permanent harm – the symptoms should spontaneously resolve when you take a break. Otherwise, spectacles can do a little of the focusing work required to look at a screen.

Many people also report that their eyesight deteriorated shortly after starting a new (screen-based) job. Invariably, this coincides either with increased reading (papers or a computer screen) or reaching middle age.

From the age of 12 or so, our ability to accommodate gradually declines as the lens inside the eye stiffens. By the early 40s, accommodation has reduced to the point where reading up close can be problematic. Those stubborn enough to persist eventually present with eyestrain.

The next question about reading from a screen is 'Does size matter?'

It probably doesn't. If the reader is able to focus on the screen (by accommodating, assisted by the correct spectacle prescription or a combination of the two), then font size won't be an issue. When not impaired by eye disease of optics, the human eye can resolve right down to phone-book sized letters and smaller.

If anything, the increased brightness of your smart phone or e-book will help you to see the fine print.

I admit that when a friend of mine suggested that I start to read on a tablet, I gave him the oft-heard response, 'I prefer the feel of books' or one of its many variants.

But I have since changed my tune, and confess to being a fully fledged e-book addict. Not only has buying books become something I can do 24/7, but I can now read in bed without annoying my wife by keeping the light on!

Men think about sex every seven seconds

Victor Minichiello, Pro Vice-Chancellor and Dean, Faculty of The Professions, University of New England
Mitra Rashidian, Postdoctoral Research Fellow, Collaborative Research Network for Mental Health, Faculty of The Professions, University of New England

Every time you turn on the television after 10 pm, eavesdrop on a group of men at your local pub, or drive past a billboard, you're likely to encounter some stereotypes about masculinity and men's sexuality.

We're told that men's minds are so immersed in thoughts of sex that it can become a full-time preoccupation. Think of James Bond's sexual exploits, Coca Cola's 'bigger is better' campaign, and the folklore that men think about sex every seven seconds (which would amount to more than 8000 thoughts about sex a day).

Let's focus, first, on one setting where there are ample opportunities for sexual interactions and discussions about sex: university. According to a 2011 study from Ohio State University, young men think about sex 19 times per day. They also have other regular, needs-based thoughts about eating and sleeping.

In contrast, the Kinsey Report, which examined the sexual behaviour of men aged under 60 years, found 54 per cent think about sex every day or several times a day, 43 per cent think about sex a few times a week or a few times a month, and 4 per cent reported just one sexual thought, or less, a month.

Another study, from 1990, found 16- to 17-year-olds think about sex every five minutes. By age 40 to 49, this drops to a sexual thought every half an hour, and it keeps reducing with age.

There's certainly no consensus among researchers about the frequency of men's sexual thoughts. And little is known about the nature of these thoughts.

So, do men think about sex more often than women?

A handful of researchers argue there are no significant differences between the frequency of men's and women's erotic thoughts outside of sex. But most studies show that men think about sex more often than their female partners. This is used to support the statement that men have more powerful sex drives than women.

Studies have suggested testosterone contributes to men's frequent preoccupation with sexual thoughts. In other words, because men have a higher level of testosterone than women, they have more frequent sexual fantasies and a stronger desire for sex.

Men's sexual fantasies tend to be more explicit than women's. And interestingly, men are more likely to fantasise during masturbation (86 per cent of the time) compared with women (69 per cent of the time).

This difference has been attributed by some researchers to men having greater opportunities – culturally and biologically – to experience sexual fantasies.

Why men think about sex

A multitude of factors could contribute to some men's preoccupation with sexual thoughts, feelings and behaviour. A 2009 study, for example, found that factors such as emotional distress, discouragement, poor self-esteem, difficulties coping with stress, and self-doubt were associated with hyper-sexuality.

Psychologist Michael Bader suggests that sexual fantasies, and resulting sexual arousal, have more to do with unconscious problem-solving than most of us realise.

But men's preoccupation with sexual thoughts cannot be fully understood without considering the effects of social media and constant internet access.

Young men are increasingly using Facebook to share pictures and stories about their sexual conquests. And the prospect of 24/7 access to pornography via mobile phones and laptops may prompt compulsive behaviour and excessive sexual thoughts.

More sex

There are other myths about the sexual character of men: they should aspire to be virile, 'well-endowed' studs and always ready for sex. But most men are not 'well endowed': the average penis size is not nine inches but, rather, between five and seven inches.

As for being ever-ready for sex, as men age, they have sex less frequently and some may even need medication to help with erectile function.

So the question we need to ask is: 'Who benefits from the perpetuation of these myths?' Perhaps Coca Cola or the sex industry. But certainly not men.

SPF50+ sunscreen almost doubles the protection of SPF30+

Ian Olver, *Clinical Professor of Oncology, Cancer Council Australia*

Australia's sunscreen regulations changed in late 2012, enabling manufacturers to label their products as SPF50+.

The sunscreen industry championed the change, instigated by Standards Australia, because the SPF50+ label could prompt many Australians to buy new products, thinking they're getting significantly higher protection from the sun.

But what does SPF50+ actually mean? And will it provide better protection?

The Sun Protection Factor (SPF) indicates the amount of UVB radiation that can reach the skin (and cause sunburn) with sunscreen, compared with no sunscreen.

In other words, SPF ratings indicate the multiples of time you could spend unprotected in the sun without burning, assuming the UV rating was constant.

But no sunscreen offers full protection from the sun. And the increment in UVB filtering between SPF30+ and SPF50+ is small, increasing protection from 96.7 per cent to 98 per cent. That's a 1.3 per cent increase, not almost double, as many people may think when making a purchasing decision.

Many sunscreens contain a combination 'inorganic' (minerals, produced using chemical processes) and 'organic' (chemical) ingredients.

Inorganic ingredients both absorb and reflect UV radiation, whereas organic ingredients only absorb. This means the energy from the UV radiation is used to convert the organic chemical into another form. But you wouldn't feel any heat produced from such a change.

As our understanding of sunscreen's role in protecting consumers from skin cancer evolves, sunscreen manufacturers are offering other protections. 'Broad spectrum' sunscreens now protect against UVB *and* UVA radiation, which we now know contributes to the development of skin cancer.

Inorganic ingredients, such as titanium dioxide and zinc oxide, may offer a broad spectrum protection but they simply reflect the UV. They also tend to be gentler on the skin.

So what are the likely results of having SPF50+ on the market?

My concern is that consumers will think the increased SPF factor offers significantly better protection than the products they're accustomed to. And if this leaves Australians using less sunscreen and neglecting other protection behaviours, we're likely to see a future spike in skin cancers.

Australia has one of the highest rates of skin cancer in the world due to our climate and large fair-skinned population. More than 10 300 Australians are diagnosed with a melanoma each year and an estimated 434 000 people are treated for one or more non-melanoma skin cancers.

Despite the popular slip, slop, slap campaign from the 1980s, more than 1830 Australians die from a skin cancer each year. Even though it's largely preventable.

Skin cancers form when skin cells are damaged by UV radiation penetrating the skin. Tanning without burning can still cause damage – if you've been exposed to enough UV to cause tanning, sufficient damage has been done to cause cancer.

It doesn't matter whether you use SPF30+ or SPF50+ sunscreen, the best way to protect yourself from skin cancer is with a combination of clothing (slip), sunscreen (slop), hat (slap), sunglasses (slide) and shade (seek), whenever the UV index reaches three or above.

Tips for applying sunscreen

- Make sure your sunscreen is at least SPF30+, water resistant and broad spectrum, which protects you from UVB and UVA.
- Apply 20 minutes before you go outdoors and reapply every two hours.
- Use at least one teaspoon of sunscreen for each limb, your face and the front and back of your body.
- Check the use-by date.

Never rely on sunscreen – whether it's SPF30+ or SPF50+ – as your only defence against the sun.

Stress can turn hair grey overnight

Michael Vagg, *Clinical Senior Lecturer, Deakin University School of Medicine; and Pain Specialist, Barwon Health*

The belief that nervous shock can cause you to go grey overnight (medically termed *canities subita*) is one of those tales which could nearly be true. There are certainly cases in medical literature of rapid greying over quite short periods of time. And reported cases go back to antiquity, including such famous figures as Thomas More and Marie Antoinette.

The biology of the phenomenon suggests that a mixture of hormones and cognitive bias is responsible for the myth.

There is little doubt that plausible biological mechanisms (Peters *et al.* 2006) exist to account for emotional stress potentially affecting hair growth. What's fascinating to me, as a pain specialist, is that several of the signalling proteins involved, such as nerve growth factor and substance P (a neuropeptide that functions as a neurotransmitter and neuromodulator) are the very same ones that operate in other nerves to create and regulate pain signals.

Human hair cycles between a growth phase (anagen), a resting phase (catagen) and a dormant phase (telogen). During the anagen phase the hair follicle produces pigment to colour the hair.

The length of the anagen phase varies according to your genes and certain hormonal levels. It can be anything between two years and eight years. When the follicle receives orders to end the anagen phase, it stops producing more hair and begins to prepare for telogen. Telogen phase lasts for between six and 18 months at a time before heading back into anagen.

After 10 or so of these cycles the follicle runs out of pigment and produces a hair with no colour at all. Despite its white colour, we insist on referring to these as 'grey hairs' for some obscure linguistic reason.

Intense stress can cause large numbers of your follicles to hit telogen at around the same time, producing simultaneous loss of a large percentage of coloured hair. This phenomenon is known as *telogen effluvium.*

Telogen effluvium is often caused by drugs which affect the hormonal control of the hair cycle, including chemotherapy drugs and anti-Parkinson's drugs.

Interestingly, these hormonal signals have a less potent effect on non-coloured hair, so a person could conceivably lose large amounts of coloured hair, leaving behind mostly white hair. This could also happen at a stressful time, such as the night before your execution. It can also happen due to auto-immunity (*alopecia areata*) where feral antibodies target pigment-producing follicles ahead of non-pigmented ones.

The problem for the myth is that none of this can happen as suddenly as overnight.

There are also plenty of good alternative explanations for these reports. In the case of Marie Antoinette, she was seen little in public in the couple of weeks before her execution, and would also have been deprived of her wigs and servants to dye her hair, if indeed that was one of her guilty secrets.

People such as President Obama, who go visibly greyer during a period of extreme stress over months or years, are usually at an age where many of their unfortunate follicles are on their last pigment cycle.

Confirmation bias means we remember those stressed people who look much greyer, but don't remember those who go through such periods without visible greying.

We also tend to ignore those who grey early and don't seem particularly stressed. That gets put down to genetics rather than stress.

So no matter how stressful your life may become, it might help to know that although you may achieve your pigmentary potential a little ahead of schedule, you can't go grey overnight.

Reference

Peters EM, Arck PC, Paus R (2006) Hair growth inhibition by psychoemotional stress: a mouse model for neural mechanisms in hair growth control. *Experimental Dermatology* **15**(1), 1–13.

Sugar makes kids hyperactive

Tim Crowe, *Associate Professor in Nutrition, Deakin University*

Any parent would tell you that seeing children fuelling up on sugar-laden cake, lollies and soft drinks at a birthday party is a sure-fire recipe for a bunch of rampaging hyperactive kids.

The connection between sugar and hyperactivity is one of the most popular food behaviour myths going around, yet it is one that has been well and truly busted by science.

Where there's sugar, there must be hyperactive kids – or so says conventional wisdom.

Science says otherwise. An abundance of published randomised controlled studies have been unable to find any difference in behaviour between children who ate sugar (from lollies, chocolate or natural sources) and those who did not.

Even studies that included children with attention-deficit/hyperactivity disorder (ADHD) could not detect any meaningful difference between the behaviour of children who ate sugar compared with those who did not.

The most important aspect of all these myth-busting studies is they used a study design where the researchers (or parents) and the children were unaware of whether they were consuming a product containing sugar or a non-sugar substitute.

It is only when you introduce an intentional bias into the experiments – and allow the parents to know what food their child was given – that the real culprit behind the myth emerges.

When parents believe their child has been given a drink containing sugar, they consistently rate their child's behaviour as more hyperactive, even if the drink did not contain any sugar.

So why do kids seem so hyperactive when they consume an abundance of sugar?

It all comes down to the context. When kids are having fun at birthday parties, on holidays, and at family celebrations, sugar-laden food is frequently served.

It's the fun, freedom and contact with other kids that makes them hyperactive, not the food they consume.

But that doesn't mean hyperactivity should be ignored. ADHD is a serious behavioural and developmental disorder that can impact on the child's academic performance and family life.

As such, extreme hyperactivity should be investigated by an appropriate health professional. Simply removing sugar from the child's diet isn't going to reduce their hyperactivity.

In fact, eliminating whole food groups in an attempt to treat hyperactivity is an extreme approach that can do more harm than good.

Growing children have different nutrient needs to adults, so eliminating whole food groups – without a valid medical reason – can affect their growth, overall health, and even later-life food preferences.

Having too much sugar, especially if it is coming from drinks, has been linked to excess weight gain and dental problems in kids. So even with the sugar equals hyperactivity myth busted, there are valid reasons to restrict how much kids consume.

Wearing tight undies will make you infertile

Robert McLachlan, *Head, Clinical Andrology,*
Prince Henry's Institute

Most men have a preference for boxers or briefs, but which kind is better when it comes to fertility?

Many things can affect a man's ability to make or transport sperm, including sexually transmitted infections, prostate and testicle infections, drug use, smoking, obesity and, perhaps surprisingly, heat.

Sperm are made in the testes (or testicles), a pair of egg-shaped glands that sit in the scrotum. It's a lengthy and continuous process. It takes about 70 days for germ cells to develop into the mature sperm, found in semen, that can fertilise an egg.

Around one in 20 men has some kind of fertility problem, with low numbers of sperm in their ejaculate. But one in every 100 men produces no sperm at all.

Of the couples who present with fertility problems, almost half are due to fertility issues in the male partner only, or in both partners. It's important to note, though, many men will still be able to father children naturally, even with a lowered sperm count.

The location of the testes in the scrotum makes the testes vulnerable to trauma, but it serves a strategic purpose – to keep them around 2°C cooler than normal body temperature, which is required for the production of top-quality sperm.

Normally, the sweating of the scrotal skin serves as an 'evaporative air cooling system' for the testes. But if it's too hot and the scrotum can't sweat, the testes will have trouble making sperm.

If the testes are too hot for too long, sperm production is interrupted and won't return to normal until their temperature returns to normal. It can take a few months of keeping the testicles at a normal temperature for sperm counts to improve if they have been lowered by heat stress.

Excess testicular heating can happen internally, as a result of feverish illnesses such as severe influenza. In cases such as these, there is no way to avoid the testes overheating because the core body temperature may be 40°C or more.

As for external testicular heating, this is more easily avoided by saying no to spas, saunas and hot baths.

While wearing tight-fitting underwear might seem to increase the temperature of the scrotum, there is no evidence to suggest it leads to infertility due to impaired sperm production.

A review of studies of genital heat stress on semen quality published in *Andrologia* (2007) found no conclusive link. The review looked into several aspects of heat stress, including more than 10 studies on wearing tight underwear and/or tight clothing.

The reviewed studies, from varied population groups and different methodologies, showed that while wearing tight-fitting underwear or clothing was associated with increased scrotal temperatures, there was no clear evidence that this resulted in reduced semen quality.

The upshot of all this is that more research is needed to answer the question of whether wearing tight underwear makes a significant contribution to fertility problems in men who would otherwise have normal sperm counts.

But for men with an already low sperm count who are trying to conceive, keeping the scrotum as cool as possible should give the testes the best chance to do their job. Keeping a pair or two of boxers on hand isn't such a bad idea.

You can selectively train your left or right brain

Annukka Lindell, *Senior Lecturer, School of Psychological Science, La Trobe University*

When it comes to things we think we ought to do, getting your body in shape often tops the list. But what about your brain?

If your left or right brain is feeling a little flabby, there's a wide range of books, teaching programs, and even a Nintendo DS game, purporting to train your left and/or right brain. Indeed, if you Google 'right brain training', you'll score 84 100 000 hits.

These products are based on the belief that the left and right hemispheres are polar opposites. The left brain is often characterised as your intelligent side: rational, logical and analytic. In contrast the right brain is stereotyped as the 'touchy-feely' hemisphere, viewed as artistic, creative and emotive.

Such left and right brain stereotypes have led theorists to suggest that people can be classified according to their 'hemisphericity'. If you're a logical, rational scientist, for instance, you're left-brained. But creative types, from artists to writers, are right-brained.

Being left-brained or right-brained often pops up in popular culture. In the business world, 'left-brainers' are complimented on their logical approach, and right-brained is synonymous with being creative/emotive.

But although the notion of 'hemisphericity' has captured the popular imagination, it is *not* supported by neuroscientific research.

Everyone, from winners of the Nobel Prize in physics to the artists behind the Archibald Prize, uses both sides of the brain when performing any task. In fact, the idea that people can be classified as left- or right-brained was debunked in scientific literature in the 1980s.

Despite this, left/right-brain training programs appear to be *gaining* popularity. This is puzzling because there's no evidence indicating that you can train just one side of your brain. Such attempts are doomed because the two hemispheres are heavily interconnected and constantly communicating.

In a normal brain, the left and right sides are connected by a band of some 250 million nerve fibres (known as the corpus callosum). And information transfer across the corpus callosum is extremely efficient.

If I show a picture to just the right brain (easily done using computer-based techniques), that information is transmitted to the left brain within 20 milliseconds (i.e. two hundredths of a second)!

The corpus callosum allows virtually instant communication between the two halves of a normal brain. This means the whole brain is involved in processing, no matter how analytic or artistic the task.

Only patients who've had their corpus callosum surgically severed can process information within just one hemisphere. This rare operation is used to relieve severe epilepsy in people who are not responding to drugs. But in a normal brain, you cannot restrict information to one hemisphere, no matter how hard you try.

New neuroscience techniques, such as Diffusion Tensor Imaging, have been specifically designed to show connections between different regions of the brain. Research using such techniques indicates that both sides of your brain are involved in everything you do.

Whether you're working on trigonometry, playing the ukulele, or taking part in 'right-brain' training, both your left and right brain are simultaneously processing and integrating information.

So try as you might, it just isn't possible for someone with a normal brain to selectively use just one hemisphere. And at present there's no independent evidence validating the claims of the programs, educational tools, and books claiming to selectively activate the right (or left) brain.

Until such evidence is available, trying to train just one side of your brain really is half-witted.

Vitamin C prevents colds

Michael Tam, Lecturer in Primary Care, University of NSW

Vitamin C is so often suggested as a treatment for the common cold that it's almost considered common sense. This well-known vitamin is primarily found in fruits and vegetables, with small quantities in some meats.

With a healthy diet, most of us should get all the vitamin C we need from food. But this doesn't stop many Australians boosting their intake through vitamin supplements.

A story on vitamin C should start with scurvy. The other name for vitamin C is ascorbic acid, which literally means 'anti-scurvy'. As vitamin C is required to build and repair body tissue, its deficiency leads to a range of horrible symptoms including bruising, bleeding, loose teeth and poor wound healing.

Until the modern era, scurvy was a major cause of death in those without access to fresh food, particularly sailors on long sea-voyages and medieval city dwellers.

The history of vitamin C tells an important story about science in medicine – and informs us about three key elements of medical research:

1. Treatments need to be tested in clinical trials

The first (documented) clinical trial of a medical treatment was by Scottish Royal Navy physician James Lind in 1747. He divided sailors suffering from scurvy into different treatment groups and found the sailors who received oranges and lemons made a dramatic recovery.

Although fresh citrus fruits had been reported as effective for scurvy before Lind's study, their efficacy had never been tested systematically. At that time, it was but one of many purported treatments (most of which we now know to be useless). Lind had demonstrated an unambiguously effective treatment for a potentially deadly condition.

2. Medical practice needs to be informed by new research

The medical establishment in the 1700s rejected Lind's findings. The prevailing view was that scurvy was related to spoiled food and poor hygiene. As this was an age where clinical trials weren't the norm, no one tried to replicate the findings.

It wasn't until four decades later that another Scottish Navy physician, Gilbert Blane, instituted health reforms and mandated the use of lemon juice.

3. Assumptions without empirical confirmation are risky

Unfortunately, scurvy was far from conquered. Fresh citrus fruits were impractical on long sea-voyages so juice and concentrates were carried instead.

In the late 1800s, a change in the preserving process and a switch to limes resulted in a juice that was devoid of any vitamin C content – useless for preventing scurvy. It was wrongly assumed that it was the 'acidity' that mattered. The lack of therapeutic testing contributed to the disastrous 1911 Scott expedition to the South Pole. The team members were beset with scurvy, 150 years after Lind's experiment.

Scurvy was finally identified as a nutritional deficiency in the early 1900s, and by the 1930s, vitamin C was found to be the essential nutrient involved.

Moving forward a few decades, the belief in the effectiveness of vitamin C for colds gained momentum following the publication of *Vitamin C and the Common Cold* by esteemed chemist Linus Pauling (one of the few people to have won more than one Nobel Prize: the second was the Peace Prize). Pauling extensively promoted vitamin C as having a wide range of health benefits and took large regular doses of supplements.

But like the history of citrus and scurvy, it's not enough to simply claim reasons why something should work. Assumptions are risky and treatments need to be tested. So what does the research evidence actually show?

When used as a treatment for cold symptoms in the general population, vitamin C supplements appear to do no better than a placebo, even in large doses (greater than one gram a day).

If you take vitamin C supplements every day for prevention, you still won't avoid any colds. But the symptoms may be milder and the duration of symptoms slightly reduced – about half a day for a typical cold lasting a week.

It's important to remember that we don't know if regular high-dose vitamin C supplements are entirely safe when taken over the long term. There is some evidence to suggest they aren't.

We all become afflicted by the common cold at times and it would be wonderful if something as simple as vitamin C supplements made a meaningful difference. But unfortunately, as the saying goes, many a beautiful theory has run aground on awkward fact.

Even taking the most favourable interpretation of the evidence, vitamin C supplements have only a minor effect on symptoms – and that's only if they're taken every day.

Wait 30 minutes after eating before you swim

Peter Milburn, *Professor, School of Rehabilitation Sciences, Griffith University*

The old saying that you should wait at least 30 minutes after eating before you swim is based on the idea that after a big meal, blood will be diverted away from your arms and legs, towards your stomach's digestive tract. And if your limbs don't get enough blood flow to function, you're at risk of drowning.

But is it sound advice, or is it just parents wanting a half-hour break to relax after a big lunch? For a fuelled-up child wanting to get back in the water, this can seem like eternity.

It's true that digestion redirects some of the blood from the muscles to aid in the digestive process. With a reduced blood flow, there is potentially less oxygen available to the working muscle and stomach, which is a potential cause of cramping – though some researchers discount this theory.

Cramps are involuntary, spasmodic contractions of skeletal muscle during or after exercise, usually related to fatigue. But cramping during exercise is more likely due to a combination of factors, such as dehydration, electrolyte imbalance and neurological fatigue, which are unique to each person.

The truth is, we have enough blood to keep all our body parts functioning after a big meal.

Another suggested risk factor for swimming after eating is what's commonly referred to as a stitch (exercise-related transient abdominal pain or ETAP in sports literature): sharp pain felt just below the rib cage. Stitches aren't well understood but are thought to be caused by cramping of the diaphragm due to restricted blood flow from pressure from the lungs above and abdomen below.

With any vigorous exercise after eating, there could be some discomfort such as heartburn or vomiting, caused by unexpected reflux or involuntary regurgitation. This is more likely to occur when there's an increase in external pressure, such as while diving.

So what do the data say about the myth?

An examination of the Royal Lifesaving Association's Australian reports on drowning over the past few years gives no mention of lives being lost after eating. And neither the American Academy of Paediatrics, the United States' Consumer Product Safety Commission, nor the American Red Cross offer any guidelines or warning related to swimming after eating.

These organisations are far more concerned with the elevated risk of drowning due to drinking alcohol. Alcohol and drugs can severely impair judgement and physical ability, and increase the risk of vocal chord spasm if water enters the windpipe.

In the 2010–11 reporting period, 17 per cent of all drownings in Australia were attributed to alcohol or drugs. In the 18–34 age group, this figure was much higher – up to 45 per cent. So it's important to be aware of the risk of alcohol and drugs when in, on and around the water.

While swimming on a full stomach can be uncomfortable and, if excessive, can lead to vomiting, it's unlikely to put you at greater risk of drowning. This will be great news for kids, but less so for their parents wanting to rest after lunch.

Common sense, however, suggests that swimming is not the best way to settle that full stomach. If you're keen to get back to the water quickly, opt for foods high in simple carbohydrates. They're not only good for you, they digest far more quickly than the fat and protein in a barbecued steak.

Mind and brain

'Biology gives you a brain. Life turns it into a mind.'

Jeffrey Eugenides, *Middlesex*

'There is a looming chasm between what your brain knows and what your mind is capable of accessing.'

David Eagleman, *Incognito: The Secret Lives of the Brain*

'How can a three-pound mass of jelly that you can hold in your palm imagine angels, contemplate the meaning of infinity, and even question its own place in the cosmos? Especially awe inspiring is the fact that any single brain, including yours, is made up of atoms that were forged in the hearts of countless, far-flung stars billions of years ago. These particles drifted for eons and light-years until gravity and change brought them together here, now. These atoms now form a conglomerate – your brain – that can not only ponder the very stars that gave it birth but can also think about its own ability to think and wonder about its own ability to wonder. With the arrival of humans, it

has been said, the universe has suddenly become conscious of itself. This, truly, it the greatest mystery of all.'

V.S. Ramachandran, *The Tell-Tale Brain: A Neuroscientist's Quest for What Makes Us Human*

'If our brains were simple enough for us to understand them, we'd be so simple that we couldn't.'

Ian Stewart, *The Collapse of Chaos: Discovering Simplicity in a Complex World*

The brain

Kate Hoy, *Research Fellow/Clinical Neuropsychologist, Monash University*

If I had been asked 15 years ago to write a short piece about what the different parts of the brain did, it would have been a fairly straightforward task. Not any more.

Over the last 15 years, the methods used to study the brain have advanced significantly, and with this so has our understanding. Which makes the task of explaining the most complex organ in the body, well, complex.

Back to basics

The structural anatomy of the brain is certainly well defined and the more basic of our functions have been generally well mapped. The 'lower levels,' such as the brainstem, regulate functions such as heart rate, breathing, and maintaining consciousness.

And the cerebellum is critical for the control and regulation of movement. While it was thought that this was its sole function, more recently the cerebellum has also been shown to have a role in so-called 'higher functions' such as cognition and emotion.

As we move to the 'higher levels' of the brain, namely the cerebral cortex, where more complex functions come into play, the assignment of function to structure becomes decidedly less distinct.

Different hemispheres

The cortex is structurally divided into two hemispheres (left and right) each with four lobes (occipital, parietal, temporal and frontal).

Brain functions, such as visual perception, language, memory, spatial ability and problem solving, have been traditionally allocated to one such lobe and/or hemisphere of the brain.

This has led to several misconceptions about brain function, the most popular of which is the commonly held belief that there is a distinction between the left 'logical' brain and the right 'creative' brain. In fact, such complex behaviours are not determined by a specific brain region, or even a specific hemisphere.

The conceptualisation of an almost one-to-one relationship between structure and function was largely a result of lesion studies, where damage to a specific part of the brain resulted in impairments in a particular function. But as our techniques of assessing the brain became more sophisticated this approach was shown to be somewhat simplistic.

We have come a long way from the phrenology of Franz Gall in the 19th century, in which characteristics such as secretiveness, self-esteem and wonder were determined by the shape of the skull (thought to be a proxy of brain size), and the 20th century reliance on lesion studies to determine the function of the different areas of the brain.

Connected network

We are now developing an understanding that complex, higher-level brain functions are a result of several brain areas working together, in what are termed 'networks'.

This has been a result of techniques such as magnetic resonance imaging (MRI), which allows us to look at the entirety of brain regions involved in certain functions, with newer applications like Diffusion Tensor Imaging allowing the visualisation of connections between these brain regions.

This is not to say that there is no separation of function throughout the brain. Rather, while there are brain regions that carry out specialised functions, they are now thought to do so in concert with other brain regions via network connections.

To conceptualise this, you could think of the brain as an exceptionally efficient rail network, where certain train stations perform specialised duties but do so in conjunction with other stations, and are connected and 'communicate' via the rail network.

Language can provide a good example of how this occurs in the brain. Language is often thought of as a solely 'left brain' function and, while there is a degree of lateralisation, this is certainly not the whole story.

There are specific regions in the dominant (usually left) hemisphere that are integral in the production and comprehension of speech, e.g. Broca's area in the frontal lobe and Wernicke's area in the temporal lobe.

But the non-dominant (usually right) hemisphere is also involved in language, and is thought to be important in the recognition and production of the emotional context of speech.

Additionally, the 'language network' involves several other dominant 'left' hemisphere regions, including the prefrontal cortex, premotor cortex, supplementary motor area, as well as regions of the parietal and temporal lobes.

These brain regions work together to perform higher order aspects of language such as the application of the correct syntax to speech, as well the mapping of words to their meaning.

While there are certain highly specialised brain regions for language, they are still part of an extensive network of brain regions which work together to produce this complex function.

In addition, the brain is not fixed in its functioning. It is plastic and, if necessary, following illness or injury, it can recruit new regions and/or networks to take over the functions of the damaged areas.

And so we believe it is a complex interaction between structure and function that best describes what the different part of the brain do – at least for now …

What are phobias?

Nick Haslam, *Professor of Psychology, University of Melbourne*

A life without fear sounds idyllic but it would be no paradise. Fear protects us from present danger, alerts us to future threat, sharpens our minds and blunts our selfishness. Friedrich Nietzsche said fear is the mother of morals, and people who lack it do indeed tend to be – as Thomas Hobbes almost put it – nasty, brutish and short-lived.

While fear can be useful to a point, people often suffer from an excess of it. Although many of us are afraid of snakes, spiders, heights and blood, when these normal fears are taken to extremes they become phobias.

To qualify as a phobia, a fear must be lasting, intense and seen by the sufferer as excessive and irrational. It must also be a source of distress or impairment in the person's occupational life and social relationships.

Phobias affect about 10 per cent of the general population at some point in their lives, with women affected twice as commonly as men.

What are we afraid of?

Phobias commonly involve objects and situations that were realistic dangers for our distant ancestors: poisonous or vicious animals and invitations to injury. As a result, many people are terrified of things that no longer pose a contemporary threat.

Ancestral fears are learnt with remarkable ease. One study found that young rhesus monkeys acquired a fear of snakes when they viewed a film of older monkeys acting terrified in the presence of a snake, but did not come to fear flowers when they viewed monkeys going ape in the presence of a blossom. Fears related to things that were threats to our forebears are more easily acquired than others.

Although many common phobias are of this ancient or 'prepared' kind, the spectrum of human fears is astonishingly broad. The clinical literature records phobias of rubber bands, dolls, clowns, balloons, onions, being laughed at, dictation, sneezing, swings, chocolate and the wicked, beady eyes of potatoes. Unusual fears are particularly common among people with autism, who have been known to dread hair dryers, egg-beaters, toilets, black television screens, buttons, hairs in the bathtub and facial moles.

It is hard to see the evolutionary threat posed by these innocuous things. As Stanley Rachman, the psychologist who treated the chocophobe, wrote, 'it is difficult to imagine our pre-technological ancestors fleeing into the bushes at the sight of a well-made truffle'.

How do phobias develop?

Given that many modern phobias make little rational sense, it is interesting to explore how they emerge. There are three main identified ways that phobias come about: a terrifying personal encounter, witnessing another person's fright, and receiving threatening information. A person might acquire a spider phobia after a close encounter in the shower, after seeing a sibling run screaming from an infested room or after being told that spider bites cause you to turn purple and die.

Only a small minority of people will develop phobias after common experiences such as these. Those who had inhibited temperaments in childhood and neurotic personalities in adulthood are more vulnerable, and this vulnerability has a substantial genetic component.

A study that followed a sample of young women over a 17-month period found that those who developed phobias tended to have more pre-existing psychological problems, poorer coping skills and a more pessimistic mindset than their peers.

Let's consider one odd but surprisingly common aversion, the fear of frogs.

One published case documented a woman who developed ranidaphobia, as it is known, after running over a knot of frogs with a lawn-mower. Overwhelmed by fear and tormented by amphibian dreams, she was persecuted every evening by an accusing chorus of survivors on a nearby riverbank.

In another case, a Ghanaian schoolboy developed his phobia when he stepped on a frog while touching leaves that irritate the skin. After his brother told him that frog urine could cause itching and a painful death, the boy became paralysed with the fear that frogs were hiding in his bed.

This fear was put to productive use elsewhere in western Africa, with one anthropologist reporting that bed-wetting children were frightened into bladder control by having a live frog attached to their waists.

What gives these puny creatures – with big eyes and scrawny, hairless bodies – their power to inspire fear and trembling? They pose no realistic threat to life: phobic individuals understand that in an encounter with a frog they are unlikely to be the one to croak.

The fear of frogs is viscerally unreasonable. To many people it reflects the frog's slimy, skin-crawling ickyness. To others, it is the creature's propensity for sudden movement, a trait it shares with another tiny source of terror, the mouse.

Treatment

Luckily for phobia sufferers, treatment is generally quick and effective. Cognitive behaviour therapists have an assortment of techniques for confronting fears and challenging the avoidance and thinking biases that sustain them. Usually these methods involve progressive exposure to the feared object or situation up the steps of a 'fear hierarchy', from relatively non-threatening encounters to the most terrifying.

These 'behavioural experiments' are often supplemented by relaxation techniques, modelling of exposure by the therapist and correction of catastrophic thoughts.

In another case of ranidaphobia, a young nursing student fainted in a biology class when her laboratory partner severed a frog's spinal cord ('pithing'). She undertook a course of therapy in which she repeatedly viewed a videotape of the operation and practised relaxation techniques.

Such was the success of the treatment that in a single sitting immediately afterwards she was able to deliver electric shocks to one frog, pith another and cut open the abdomen of an anaesthe-tised rat, remaining calm even when one frog hopped loose, bleeding profusely from its injuries.

By facing what we dread, under the guidance of a psycholo-gist, we can find freedom from irrational fear.

What is déjà vu and why does it happen?

Amy Reichelt, Research Fellow in Neuroscience, University of NSW

Have you ever experienced a sudden feeling of familiarity while in a completely new place? Or the feeling you've had the exact same conversation with someone before?

This feeling of familiarity is, of course, known as déjà vu (a French term meaning 'already seen') and it's reported to occur on an occasional basis in 60 to 80 per cent of people. It's an experience that's almost always fleeting and it occurs at random.

So what is responsible for these feelings of familiarity?

Despite coverage in popular culture – from *The Matrix* to Monty Python – experiences of déjà vu are poorly understood in scientific terms. Déjà vu occurs briefly, without warning and has no physical manifestations other than the announcement: 'I just had déjà vu!'

Many researchers propose that the phenomenon is a memory-based experience and assume the memory centres of the brain are responsible for it.

Memory systems

The medial temporal lobes are vital for the retention of long-term memories of events and facts. Certain regions of the medial temporal lobes are important in the detection of familiarity, or recognition, as opposed to the detailed recollection of specific events.

It has been proposed that familiarity detection depends on rhinal cortex function, whereas detailed recollection is linked to the hippocampus.

The randomness of déjà vu experiences in healthy individuals makes it difficult to study in an empirical manner. Any such research is reliant on self-reporting from the people involved.

Glitches in the matrix

A subset of people with epilepsy consistently experience déjà vu at the onset of a seizure – that is, when seizures begin in the medial temporal lobe. This has given researchers a more experimentally controlled way of studying déjà vu.

Epileptic seizures are evoked by alterations in electrical activity in neurons within focal regions of the brain. This dysfunctional neuronal activity can spread across the whole brain like the shock waves generated from an earthquake. The brain regions in which this electrical activation can occur include the medial temporal lobes.

Electrical disturbance of this neural system generates an aura (a warning of sorts, not necessarily visual) of déjà vu before the epileptic event.

By measuring neuronal discharges in the brains of these patients, scientists have been able to identify the regions of the brain where déjà vu signals begin.

It has been found that déjà vu is more readily induced in epilepsy patients through electrical stimulation of the rhinal cortices as opposed to the hippocampus. These observations led to the speculation that déjà vu is caused by a dysfunctional electrical discharge in the brain.

These neuronal discharges can occur in a non-pathological manner in people without epilepsy. An example of this is a hypnagogic jerk, the involuntary twitch that can occur just as you are falling asleep.

It has been proposed that déjà vu could be triggered by a similar neurological discharge, resulting in a strange sense of familiarity.

Some researchers argue that the type of déjà vu experienced by temporal lobe epilepsy patients is different from typical déjà vu.

The déjà vu experienced before an epileptic seizure may be enduring, rather than a fleeting feeling in those who don't have epileptic seizures. In people without epilepsy the vivid recognition combined with the knowledge that the environment is truly novel intrinsically underpins the experience of déjà vu.

Mismatches and short circuits

Déjà vu in healthy participants is reported as a memory error which may expose the nature of the memory system. Some researchers speculate that déjà vu occurs because of a discrepancy in memory systems, leading to the inappropriate generation of a detailed memory from a new sensory experience.

That is, information bypasses short-term memory and instead reaches long-term memory.

This implies déjà vu is evoked by a mismatch between the sensory input and memory-recalling output. This explains why a new experience can feel familiar, but not as tangible as a fully recalled memory.

Other theories suggest activation of the rhinal neural system, involved in the detection of familiarity, occurs without activation of the recollection system within the hippocampus. This leads to the feeling of recognition without specific details.

Related to this theory, it was proposed that déjà vu is a reaction of the brain's memory systems to a familiar experience. This experience is known to be novel, but has many recognisable elements, albeit in a slightly different setting. An example? Being in a bar or restaurant in a foreign country that has the same layout as one you go to regularly at home.

Even more theories exist regarding the cause of déjà vu. These span from the paranormal – past lives, alien abduction and precognitive dreams – to memories formed from experiences that are not first-hand (such as scenes in movies).

So far there is no simple explanation why déjà vu occurs, but advances in neuroimaging techniques may aid our understanding of memory and the tricks our minds seem to play on us.

What is depression?

Philip Batterham, Research Fellow, Centre for Mental Health Research, ANU
Amelia Gulliver, Research Assistant, Centre for Mental Health Research, ANU
Lou Farrer, Postdoctoral Research Fellow and Registered Psychologist, Centre for Mental Health Research, ANU

Many people know what it's like to feel sad or down from time to time. We can experience negative emotions due to many things – a bad day at work, a relationship break-up, a sad film, or just getting out of bed on the 'wrong side'. Sometimes we even say that we're feeling a bit 'depressed'. But what does that mean, and how can we tell when it's not just a feeling?

Depression is more than the experience of sadness or stress. A depressive episode is defined as a period of two weeks or longer where the individual experiences persistent feelings of sadness or loss of pleasure, coupled with a range of other physical and psychological symptoms including fatigue, changes in sleep or appetite, feelings of guilt or worthlessness, difficulty concentrating or thoughts of death.

To be diagnosed with major depressive disorder, individuals must experience at least one depressive episode that disrupts their work, social or home life.

Depression is common in the community, with the Australian Bureau of Statistics reporting that 12 per cent of Australians experience a major depressive disorder in their lifetime. More than 650 000 Australians have this experience in any 12-month period.

Because depression is highly prevalent and can be significantly disabling, the World Health Organization reports that it is the third-highest cause of disease burden worldwide, with a greater burden on the community than heart disease. There are also high levels of overlap between depression and other common mental disorders, including anxiety and substance use disorders.

Unfortunately, only 35 per cent of people with symptoms of mental health problems seek help. This may be a result of difficulties identifying depression in the community because of a lack of knowledge, difficulty accessing care, and stigmatising attitudes towards depression.

Depression prevention programs that provide accessible treatments, increase knowledge and change negative attitudes are an important way to increase access to treatment and reduce the burden of depression.

Causes and risk factors

There's generally no single reason an individual becomes depressed. There's a constellation of risk factors, including physiological, genetic, psychological, social and demographic influences.

Biological risk factors include having a family history of depression, suffering a long-term physical illness or injury, experiencing chronic pain, using illicit drugs or certain prescription medications, chronic sleep problems, or having a baby. Having experienced depression in the past is a risk factor for a further depressive episode.

Psychological risk factors for depression include having low self-esteem or having a tendency to be self-critical. Demographic and social influences include being female (women are almost twice as likely to suffer from depression than men), stressful life events (such as relationship conflict or caring for someone with an illness), experiencing a difficult or abusive childhood, or being unemployed.

People differ greatly in the amount or type of risk factors they're exposed to or experience. And having several risk factors alone is not enough to trigger depression.

A combination of risk factors and the experience of stressful or adverse life events may prompt the onset of depression. The greater the number of risk factors a person experiences, the more vulnerable they are to developing depression when stressful life events occur.

In contrast, those exposed to fewer risk factors are somewhat buffered, and may only develop depression when exposed to extreme levels of environmental stress.

Treatment and prevention

There are several effective treatments for depression. The most effective and widely used are cognitive behavioural therapy and antidepressant medications.

Cognitive behavioural therapy is a talking therapy that primarily aims to reduce negative thinking patterns, while antidepressant medications target brain chemicals thought to be implicated in depression.

There's also evidence that low-intensity cognitive behavioural therapy combined with education about depression can prevent individuals from developing depression. To widen the reach of such prevention programs, internet therapy programs have been developed and shown to be effective in preventing depression. You can find a list at https://beacon.anu.edu.au/service/website/browse/1/Depression. Australian researchers are at the forefront of developing e-mental health platforms to reduce the prevalence of depression and other mental disorders.

There is some evidence that lifestyle changes can also help to prevent depression in some people. Engaging in healthy behaviours, such as getting adequate sleep, avoiding substance use, taking vitamins or fish oil supplements, engaging in physical activity and a healthy diet, have all been shown to have associations with reduced depression symptoms. But research continues to examine whether making changes in these areas can lead directly to the prevention of depression.

Future research

There are several promising research areas currently being explored. Researchers are investigating ways to make cognitive behavioural therapy more effective through better understanding of the processes involved in recovery. And technology has improved the availability of online, mobile and computer-based treatments, so people at risk of depression in under-served areas such as rural locations or developing countries can access evidence-based services.

Population-based research is leading to a better understanding of risk factors for depression and improvement in its early detection. Research on the biological and genetic bases of depression is resulting in continual refinement of physical and pharmacological treatments.

A more nuanced understanding of the treatment options that work best for specific individuals has great promise for allowing an individually tailored approach to treating and preventing depression.

What is dreaming?

Drew Dawson, *Director, Appleton Institute, Central Queensland University*

For most of human history, dreaming has been seen as a second 'reality' in which altered forms of perception provide insights into ourselves and others, our fears, fantasies and motivations or even the future.

What Freud referred to as the 'royal road to the unconscious' served as a source of wonderment and prophecy. So what do we think about it now?

What is dreaming? What does science say? And what mysteries remain?

In the developed world, the cultural importance of dreaming has diminished significantly over the last 100 years. In part, this reflects the increasing dominance of science in the way we understand human experience.

With the rise of neuroscience and empirically based psychologies, the dream has become increasingly irrelevant to our current understanding of brain function.

Ironically, the initial shift to a modernist interpretation of dreams started with Freud, their modern champion.

A spiritual vehicle

Animist philosophies such as in Shintoism and Native American spirituality often used dreams to partition the mundane aspects of our lives from the spiritual.

The dream was a vehicle by which they could access the animal 'spirits' often associated with the totemic affiliations and custodianship required by tribe or family group.

In theist cultures, which believe in one or more gods, the dream was more likely to be viewed as a form of divination and a medium through which god(s) could, albeit symbolically, communicate their knowledge of culture, purpose or the future.

With the decline of psychoanalysis as a cultural trope, the dream was increasingly subjected to the harsh light of science.

In the late 1950s, sleep researchers Aserinsky and Dement first identified the characteristic rapid eye movements (REM) and brain waves that enabled us to tell when an individual was dreaming.

For the next 20 years there was an incredible flourishing of dreaming research.

Medical and psychological researchers were able to wake people while dreaming (or not) and ask them about the thoughts, feelings and ideas associated with the dream state.

Such studies confirmed many long-held beliefs about dreaming. Dreaming was very similar to waking, at least in terms of the brain waves recorded on electroencephalogram (EEG) machines, which measure and record the electrical activity from different parts of the brain.

And yet dreaming consciousness was very different to waking. It was more visual; and ideation (the creation of ideas) was more bizarre and often illogically connected.

Dreaming consciousness often mixed mundane aspects of our previous waking life with strange and symbolic mental activity.

People woken from dreaming reported feelings and emotions that were quite different from those reported when in deep sleep.

People who were in deep sleep used fewer words, were less coherent in their speech patterns and were less 'conscious' than those who were awakened from dreaming sleep.

From the late 1970s until the early 2000s, dream research shrivelled to a small field. Many regarded it as a quaint anachronism marking the transition between psychoanalytic and neuroscientific conceptions of mind.

But in recent years the role of dreams in cognition has been reinvigorated by the discovery that the two basic modes of sleep – dream (REM) sleep and slow wave (deep) sleep (SWS) – play quite different roles in how we recover from the trials and tribulations of wakefulness.

In simple terms, SWS regulates physical recovery and REM mental recovery.

Initial rodent studies showed depriving animals of REM sleep was associated with impaired learning. The way in which memories are laid down and learning consolidated is profoundly linked to brain activity during dreaming sleep.

More recently, the same phenomena have been observed in human studies – and these have spawned a whole new field of REM sleep research linking the quality and quantity of dream sleep to memory and learning.

Back to the start

The story might yet come full circle. While the first generation of 'scientific' dream research did not find a simple link between the reported content of the dream and psychological health, the next generation of dream research may well uncover a link, however subtle.

Many of the drugs we use to treat depression have profound effects on REM or dreaming sleep. We know the ways in which depressed patients learn and recall memories are very different from those of people who are not depressed.

Depressed people are more likely to recall negative events, experiences and emotions, and more likely to forget positive ones. We know that people who do not get enough sleep, especially REM sleep, do not learn as effectively.

The next 20 years promise a very new and exciting period for research into REM sleep.

But if we stand aside from the immediacy of the new technologies of sleep and the 'science' of recent dream research we can see some broader patterns repeating in the human history of dreaming.

We are still looking at dreams as a different state of consciousness that merges aspects of sleep and wakefulness. We still see dreams as an aspect of mind and brain that can influence how we see and interpret the world.

We now have sufficient knowledge of genetics to see that our brains carry the seeds of the past and that the ways our brains operate reflect the collective unconscious – an idea posited by Freud's famous student, Carl Jung.

We still see dreams as a source of inspiration and a canvas upon which we can create new and different possibilities, new futures.

One can only wonder how we might understand and use our dreams in another thousand years.

What is forgetting?

Jee Hyun Kim, *DECRA Fellow, Behavioural Neuroscience, Florey Institute of Neuroscience and Mental Health*

If memory can be defined as 'a past that becomes a part of me', can forgetting be defined as 'a past that is no longer a part of me'?

Smokers who have abstained for years may not consciously be able to recall the sensation brought forth by smoking, but can suddenly feel craving upon seeing a smoking-related cue – often a cigarette brand logo – and relapse into smoking again. Tobacco companies know this only too well.

This illustrates a past that may have been forgotten but is not gone. It's consistent with what's known as the cue-dependent theory of forgetting, which states that there can be difficulty in recollection when the stimuli present during memory encoding are absent. Upon presentation of such stimuli, recollection becomes easy.

When a memory exists but cannot be recalled, such forgetting represents a retrieval failure. When a memory cannot be recovered in any way it represents a storage failure.

I have experienced such storage failure repeatedly throughout my education. Information I have studied before an exam usually lasts throughout the exam.

As soon as the exam finishes, the information seems to immediately fade, mostly beyond recovery. It's always made me feel like a faker.

This type of forgetting is consistent with what's known as the trace decay theory of forgetting. It states that without rehearsal, memory will gradually decay over time, to disappear forever.

Unfortunately, both psychological theories outlined above are very limited in explaining forgetting.

Cue-dependent theory is criticised because memories can generalise over time, their elements becoming less specific. Trace decay theory fails to explain different fading speeds of different memories.

Infantile amnesia

Neuroscience comes to the rescue at this point. What happens to the brain when something is forgotten? Although there are not many studies, some insight on the neural basis of forgetting has been provided by infantile amnesia research.

Infantile amnesia commonly refers to the general inability to remember experiences that happened early in life, before three to five years of age. This is a pervasive phenomenon displayed by all humans.

Even Dr Sheldon Cooper of TV's *The Big Bang Theory* – who has photographic, or eidetic, memory – cannot recall events before 'that drizzly Tuesday' when his mother stopped breastfeeding him.

Importantly, infantile amnesia is not due to an inability to form episodic memories before that age. Children younger than three years have been shown to be able to encode specific episodic memories and even remember them for two years.

Instead, these memories do not persist into later childhood and adulthood, indicating children forget at a more rapid rate than adults.

Amazingly, infantile amnesia may be ubiquitous. It is believed to occur in all altricial species (species that, like us, require parental care after birth), and has been observed even in worms, goldfish, chickens and rats. These species require caregivers to survive into adulthood.

This is in contrast to precocial species (species that don't require parental care) such as guinea pigs, that do not require caregiving to survive.

Rats!

In 1962, Byron A. Campbell and Enid Campbell showed that juvenile rats forget considerably faster than adult rats. They trained rats of various ages to avoid the black chamber of a black–white shuttle box.

This memory retrieval was measured by the length of time they took to enter the black chamber when placed in the white chamber. When tested immediately after training, rats of all ages were equally able to remember to avoid the black chamber.

But when testing occurred later, infant rats showed nearly complete forgetting after seven days, whereas adult rats showed nearly perfect avoidance of the black chamber even after 42 days.

This finding has since been replicated with different learning paradigms and different species, including humans.

Since that time, infantile amnesia has been placed in a 'too hard basket' and has not been studied, although most psychologists believed it was a very important clue to how our memory works.

GABA

In 2006, more than 40 years after the original study, my colleagues and I showed that a neurotransmitter called gamma-aminobutyric acid (GABA) is involved in infantile amnesia.

Reducing GABA allowed juvenile rats to retrieve a forgotten fear memory. In the mammalian brain, GABA is the king of inhibitory communication. Removing this inhibition removed whatever was blocking the fear memory from being retrieved.

Freud was right: infantile amnesia involves repression of aversive memories that can reappear later in life!

But this recovery of memory in rats was only possible 10 days after the memory was made, and reducing GABA had no effects after 60 days. So some time between 10 to 60 days in these rats, roughly equivalent to between eight months to five years for us, forgetting turned from being a retrieval failure to storage failure.

In the latest follow-up to this study, my colleagues and I showed that a brain region called the amygdala, a very primitive part of the brain responsible for emotion, expressed traces of fear memory that the juvenile rat had forgotten.

That trace was in the form of increased activation of a chemical called mitogen-activated protein kinase in the neurons, an important kinase (an enzyme involved in the transfer of energy within cells) necessary for gene activation and transcription.

Perhaps this is part of the 'engram' (a presumed means by which memory traces are stored as biochemical changes in the brain, allowing retrieval of forgotten memories) scientists have been hunting for decades.

Understanding infantile amnesia may provide a key to unlocking the secrets of forgetting that some happy individuals appear to possess. After all, people who develop a post-traumatic stress disorder following a traumatic event suffer from lack of forgetting.

And there are still so many other questions. Is memory erasure possible? What about reconstruction of memories? Will Jason Bourne ever completely recover from his amnesia caused by physical trauma?

What is post-traumatic stress disorder?

Mark Creamer, Professorial Fellow, Department of Psychiatry, University of Melbourne

People have probably always known about the psychological effects of experiencing life-threatening events such as military combat, natural disasters or violent assault. Literature through the ages – some of it thousands of years old – provides many vivid portrayals of these internal struggles to recover from horrific experiences.

It was not until 1980, however, that these reactions were formally recognised by the international psychiatric community, in the *Diagnostic and Statistical Manual of Mental Disorders*, 3rd edition. The name chosen was post-traumatic stress disorder, or PTSD, and the diagnostic criteria were agreed.

Before discussing the nature and treatment of PTSD, it's important to emphasise that human beings are generally resilient. Most people exposed to potentially traumatic events recover well with help from family and friends, and don't develop mental health problems.

For those who don't recover so well, PTSD is only one possibility, with depression, substance abuse, anxiety and physical health problems also common. But PTSD is the only condition specifically tied to a traumatic experience.

Symptoms
PTSD is a serious psychiatric disorder characterised by three groups of symptoms:

- Reliving the traumatic event. People with PTSD describe vivid, painful images and terrifying nightmares of their experience.
- Avoidance. People with PTSD try to avoid reminders of what happened. They become emotionally numb and socially isolated to protect themselves from the pain.
- Being constantly tense and jumpy, always on the look-out for signs of danger.

PTSD is associated with significant impairment in social and occupational functioning.

Prevalence and risk factors

The most recent Australian National Mental Health Survey (2007) reported that over 4 per cent of the population had experienced the symptoms of PTSD in the previous year.

The incidence of PTSD varies considerably depending on the type of trauma, with sexual assault consistently the highest (around half of rape victims will develop PTSD). Accidents and natural disasters – events that do not involve human malevolence – tend to be the lowest at around 10 per cent.

About half the people who develop PTSD recover over the first six to 12 months. Unfortunately, in the absence of treatment, the others are likely to experience chronic problems that may persist for decades.

So why do some people develop these problems and not others? The answer is a combination of what the person was like before the trauma, their experiences at the time, and what has happened since.

In terms of pre-trauma factors, genetic vulnerability plays a part, along with a history of trauma, particularly in childhood, as well as tendencies towards anxiety and depression. Not surprisingly, the more severe the traumatic experience (the higher the threat to life or exposure to the suffering of others) the more likely the person is to develop PTSD.

The final group of risk factors appears after the event, with the most important being social support: people who have a strong network of friends and family to support them after the experience are less likely to develop PTSD. Other life stressors during this period (such as financial, legal, health or relationship problems) can also interfere with recovery.

Treatment

We have come a long way in improving treatments for PTSD and now have a large body of research evidence to guide our decisions.

The most effective treatment is trauma-focused psychological therapy. There are a few different forms, including cognitive behavioural therapies, as well as something called eye movement desensitisation and reprocessing. The thing they share in common is providing the survivor with an opportunity to confront the painful memories, and to 'work through' the experience in a safe and controlled environment. This therapy is not easy for either the patient or the therapist, but a considerable body of evidence indicates it is very effective in most cases.

Pharmacological treatment can also be useful in PTSD, especially in more complex cases and as an adjunct to trauma-focused psychological therapy. The most effective drugs for PTSD are the new generation anti-depressants – the selective serotonin re-uptake inhibitors or SSRIs. Other drugs can also be useful, depending on the clinical presentation.

The bottom line is that effective treatment is available if the PTSD sufferer can find their way to an experienced clinician.

We've come a long way in our understanding of mental health responses to trauma in the last couple of decades, but many challenges lie ahead: Can we prevent the development of these problems? How should we respond with whole communities following widespread disaster such as bushfires, floods or terrorism? And can we improve the quality and availability of treatment?

As we address these challenges, we must strive to make sure the best possible care is available to those whose lives have been devastated by the experience of severe trauma.

Research and technology

'Research is the process of going up alleys to see if they are blind.'

Marston Bates, *The Nature of Natural History*

'I am busy just now again on Electro-Magnetism and think I have got hold of a good thing but can't say; it may be a weed instead of a fish that after all my labour I may at last pull up.'

Michael Faraday

'If politics is the art of the possible, research is surely the art of the soluble. Both are immensely practical-minded affairs.'

Sir Peter Medawar, *The Art of the Soluble*

'For a successful technology, reality must take precedence over public relations, for Nature cannot be fooled.'

Richard P. Feynman

Nanotechnology and you

Michael J. Biercuk, *Senior Lecturer, School of Physics, University of Sydney*

For the public, the jury is still out on nanotechnology – the media simultaneously extol its promise and warn of the potential calamity facing humanity.

But what is it? How does it work? Is it dangerous?

What we now call nanotechnology has the aim of designing, controlling and probing matter at the nanoscale.

The 'nano' bit refers, of course, to the size scale of nanometres, with one nanometre being one billionth of a metre. For scale, in a typical solid material, one can fit just 10 atoms into a nanometre.

While no single individual invented the field, much credit goes to the US physicist Richard Feynman. In a famous address to the American Physical Society entitled 'There's Plenty of Room at the Bottom', he challenged the scientific community by presenting his vision of the future of technology.

Feynman envisioned building a transistor – the core element in your computer – atom by atom, using tiny machines to build the device from the bottom up. The key concept in his vision was that of taking control of the process, instead of simply mixing things together and letting nature run the show.

Of course, various scientific disciplines such as chemistry and biology have for decades worked with matter at the atomic scale – chemical reactions take place between individual atoms, and biomolecules are studied on comparable scales. But what Feynman was describing was fundamentally different.

By emphasising the access to and control over individual systems at the scale of nanometres, Feynman realised the possibility of unlocking totally new technical capabilities.

In the scientific sense, size really matters; matter at the nanoscale behaves in a fundamentally different way than it does at the macroscale.

Where we are

Nanotechnology has developed into an extremely broad, and even loosely defined, field. It encompasses research in electronics, biological systems, chemistry, materials science and precision metrology.

We produce new materials at the nanoscale – nanoparticles and nanomaterials; engineer new active nanoscale electronic and optical elements – nanodevices; and develop new techniques to probe matter on the nanoscale – nanometrology.

Today, Feynman's vision has been realised. We have transistors made of just a single molecule, quantum devices using individually controlled trapped atoms, and an amazing array of novel nanostructured materials in laboratories all over the world.

What this means for you

A wide variety of consumer products already benefit from nanotechnology.

Believe it or not, new techniques in nanomaterials growth and nanoelectronics currently appear in the microprocessor powering your computer.

Tiny transistors with insulators just one nanometre thick and lengths of just about 20 nanometres are currently in production. Leveraging these tiny size scales allows incredibly large numbers of devices to be integrated on a chip, and enables them to operate faster than their predecessors.

Then, of course, we have sunscreen. By engineering matter at the atomic scale we can produce nanoparticle forms of standard chemicals, say titanium dioxide, which absorb ultraviolet light but are invisible in the range covered by human vision.

That way, when you slather on the sunblock you don't look grey. Other cosmetics and industrial products benefit similarly from the incorporation of nanoparticles.

Grey Goo

We've certainly been hearing about the dangers of nanotechnology – from 'Grey Goo' to toxic nanoparticles.

A word to the wise: Grey Goo, the concept that tiny nanomachines will overrun the earth, is science fiction, with the emphasis on fiction.

The existence of nanotechnology in your computer's microprocessor clearly doesn't pose a threat of self-replicating nanobots, nor does research on new electronic or optical devices allowing scientists to study unique features of quantum mechanics.

Genuine concerns

That said, there *are* real health and safety issues. Taking the example of nanoparticles in sunscreen, the trade-off for making the particles invisible by reducing their size is that they become more readily absorbed into your body.

Tests have demonstrated that depending on size and other characteristics, nanoparticles may be easily transmitted through cell membranes, or even pass the blood–brain barrier.

In itself, passage of nanoparticles through the body is not necessarily a problem – in fact it is a major motivation for the development of nanotechnologies in medical science.

Unfortunately, materials that are known to be safe for human exposure can have fundamentally different chemical and toxicological characteristics when reduced to the nanoscale. Remember, size matters.

Studies have suggested that nanoparticle aggregation in various parts of the body can be hazardous to human health.

Nanotubes, for instance (imagine tiny drinking straws just one nanometre wide), have been shown to aggregate in the lungs if inhaled, producing symptoms in laboratory animals similar to those created by exposure to asbestos. But other nanoparticles have been shown to be harmless, greatly complicating the issue.

The concerns are very real and require careful studies of risk.

Keep calm

A bit of context is always helpful. It may surprise you, but most significant sources of nanoparticle exposure are often from natural phenomena. When you go to the beach, or there's a dust storm, you inhale large quantities of silica nanoparticles (glass).

Bushfires release large quantities of nanoparticles due to the combustion process. And perhaps most surprisingly, the milk we drink every day consists of a nanoscale colloidal suspension – little blobs of fat suspended in water.

And if you're really concerned about man-made sources of nanoparticles you need to look at much less exotic technologies.

Perhaps the largest source of man-made nanoparticles to which we are exposed is the internal combustion engine. And we know these nanoparticles are toxic.

As with all technological developments, we need to be careful, but we must fight the urge to overreact. A blanket moratorium on the use or study of nanotechnology, as advocated by some environmental groups, fails to acknowledge basic science.

So moving forward, the scientific community, environmental groups, governments, industry and the public should cooperate to produce workable and useful regulations.

Fact and science always need to form the basis of our decisions.

Quadruple-helix DNA

Friederike M. Mansfeld, Research Fellow, Monash University

DNA has been called many things: the king of molecules, the blueprint of life, and less excitingly but perhaps more accurately, the genetic code.

DNA's double helix, discovered in 1953 by James Watson and Francis Crick, with contributions from Maurice Wilkins and Rosalind Franklin, is one of the most recognisable natural structures ever reported by scientists.

And recently, nearly 60 years after Watson and Crick's famous discovery, a team of Cambridge researchers has confirmed the existence, in humans, of an alternative DNA structure: a G-quadruplex.

To understand the significance of this discovery we need to understand a little about DNA itself.

Simple complexity

The simplicity of DNA's components and the spectacular complexity of the resulting organisms never cease to inspire a sense of wonder. All that's needed is just four 'bases', arranged into pairs – adenine and thymine, cytosine and guanine – and connected to two backbones wound tightly around each other.

Only the number and order of the bases determines the difference between single-celled bacteria and human life.

Although the double helix is the most common and best known structure of DNA, this fascinating molecule is capable of adopting many other structures, including a four-stranded helix.

G-quadruplex

Instead of pairs, sequences rich in guanine can form quartets consisting exclusively of this base. Stacks of these guanine quartets form a quadruple helix, also known as a G-quadruplex.

This discovery was made less than 10 years after the publication of the double helix structure but, until very recently, the existence of G-quadruplexes in human cells had not been proven.

The fact that human DNA contains many guanine-rich sequences, and that these form G-quadruplexes in a test tube, was considered to be indirect evidence for their existence in human cells.

G-quadruplexes in humans

The human genome contains about 376 000 sequences with the potential to form G-quadruplexes. Their locations in the genome suggest they are more than a mere structural curiosity and fulfil important functions in the cell cycle.

On January 20, a team of researchers at the University of Cambridge, led by Shankar Balasubramian, published a paper in *Nature Chemistry* showing strong evidence for the existence of G-quadruplexes in human cells and their involvement in the replication of cells (Biffi *et al.* 2013).

To achieve this, the researchers produced antibodies that specifically recognise G-quadruplexes but no other DNA structures. These antibodies were labelled with a fluorescent molecule that can be observed in a microscope, and used this to track the formation and location of G-quadruplex structures during the cell cycle – the series of events that take place in a cell leading to its division and duplication.

The highest number of G-quadruplexes was observed in the S phase of the cell cycle – the point at which DNA is separated into two single strands and copied before cell division. This observation suggests G-quadruplexes are important for cellular replication. The research also indicates that quadruplexes are more likely to occur in genes of cells that are rapidly dividing, such as cancer cells.

Exciting implications

This research also shows that the number of G-quadruplex structures present in human cells can be increased by a small molecule called pyridostatin, which can stabilise G-quadruplexes in the test tube.

This same molecule can also stop the growth of human cancer cells by damaging parts of the DNA that are important for replication.

There is still a long way to go towards unravelling the precise function of these quadruple helices but the findings discussed here show a link between these structures and cellular replication.

Further research to gain a better understanding of this link may eventually lead to insights into the mechanisms that cause cancer and improved treatments.

Reference

Biffi G, Tannahill D, McCafferty J, Balasubramanian S (2013) Quantitative visualization of DNA G-quadruplex structures in human cells. *Nature Chemistry* **5**, 182–186.

The 2012 Nobel Prize in Chemistry – what are G-protein-coupled receptors?

Wayne Leifert, *Senior Research Scientist, Nutritional Genomics and DNA Damage, CSIRO*

Two US scientists have been awarded the 2012 Nobel Prize in Chemistry for discovering the receptors that transmit signals such as light, taste or smell to cells.

Robert Lefkowitz (of Duke University Medical Centre in Durham, North Carolina) and Brian Kobilka (of Stanford University School of Medicine in Palo Alto, California) received the prize for their work on 'cells and sensibility'.

For a long time, it remained a mystery how all the cells in our body could specifically sense and respond to their environment. No-one really knew how these sensors could transmit an external signal across the cell membrane to the inside of the cell, resulting in a broad range of physiological processes.

When Dr Kobilka was introduced to receptors while doing postdoctoral research in Dr Lefkowitz's laboratory, their work focused on sensors located in the heart that respond to the hormone adrenaline.

These were found to be located right at the cell surface and are part of a much bigger family of receptors called G-protein-coupled receptors (GPCRs).

Thanks to the pioneering work of Lefkowitz and Kobilka, it is now known that GPCRs can detect many other internal signals which are relayed via these receptors to the inside of our cells as a message to do something unique.

For instance, when adrenaline binds to receptors in your heart (which the Nobel Prize winners characterised) all the cells may contract harder and faster. You may well have experienced this 'internal' sensation when you have become nervous about something.

Indeed GPCRs are responsible for the anxiety you may experience when a spider falls out of your sun visor when you are driving your car!

GPCRs are also used to relay messages from the external environment to provide us information. We now know there are receptors in our nose to detect odours, in our eyes to detect light and in our mouth to detect flavours. These proteins function as a kind of gatekeeper, responding to outside stimuli and transmitting them as internal biological effects.

So when you drink your favourite glass of red wine it is your receptors for odour and flavours that give you the taste sensation. And, of course, light receptors allow you to see the nice colour of your shiraz.

There are many different GPCRs located right throughout our bodies involved in everything ranging from fat metabolism to neurotransmission.

In fact most physiological processes depend of GPCRs and approximately half of all medications act on these receptors to help treat many different medical conditions.

So-called beta blockers are used to treat high blood pressure (hypertension) and antihistamines are used for allergies. Even honey bees use their GPCRs to find their way to the next flower!

Identifying and understanding how GPCRs work has been crucial to unravelling the complex network of signalling between cells and organs and the environment.

The groundbreaking work of Lefkowitz and Kobilka has opened many paths of discovery at the biology/chemistry interface and has considerable broader impact in our day-to-day lives.

More recently, researchers have begun to look at how these receptors can be regulated by drugs and by proteins other than their G protein partners. Research has also focused on what molecular mechanisms determine how a stimulant on the outside of the cell can interact with GPCRs to cause changes that are transmitted to the inside of the cell. Lefkowitz made the major contributions to the first question, Kobilka to the latter.

The most recent breakthrough came in 2011 when Kobilka and his co-workers identified the complex crystal structure involved and produced what is essentially a snap-shot of the activation process at the molecular level. The Nobel Committee described this as a 'molecular masterpiece'.

Because GPCRs are involved in virtually every biological process and most diseases – including cardiovascular disease, obesity and diabetes, neuropsychiatric disorders, inflammation and cancers – they are likely to be a fertile ground for chemistry and novel drug discovery. However, this presents a challenge: we don't yet have detailed information on how these proteins work and how they can be targeted in a way that avoids side-effects.

For example, how can we activate or block a GPCR for dopamine in one part of the brain without hitting another GPCR that also recognises dopamine in other parts of the body? And why do some chemicals that act at the same GPCR prove clinically effective, but other ones fail?

Understanding how these nanoscale molecular machines are intricately built and how they function has been, and will continue to be, crucial for the field of molecular biology in the future.

What is GPS?

Andrew Dempster, *Professor, School of Electrical Engineering and Telecommunications, University of NSW*

It's a device used widely in cars, on smartphones and in fitness devices. But what exactly is GPS, and how is it able to pinpoint our exact location anywhere on Earth?

How does it work?

The Global Positioning System (GPS) is a constellation of up to 32 satellites that orbit at a height of 26 600 km above Earth. The satellites are owned by the US Department of Defence, but anyone with a receiver can use the signals from them.

For the receiver to work, it needs to be able to 'see' four of the satellites. When you turn on your receiver, it may take a minute or so to locate these satellite signals, then download data from the satellite before positioning can commence.

Fundamentally, two things need to happen for this to work effectively:

1. The GPS receiver measures the distance from itself to a satellite by measuring the time a signal takes to travel that distance at the speed of light.

2. When the satellite's position is known, the GPS receiver knows it must lie on a sphere that has the radius of this measured distance with the satellite at its centre. The receiver need only intersect three such spheres, as seen in the image below.

This process, known as trilateration, is an effective means of determining absolute or relative locations.

Trilateration

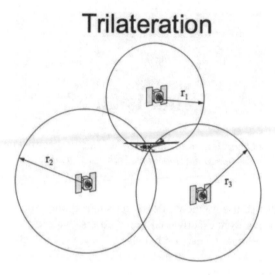

But there's a problem. GPS satellites have very expensive atomic clocks on board, accurate to one second in 32 000 years – and therefore know what time their signals are transmitted – but the GPS receiver has a very cheap clock. That means there is uncertainty about the 'receive' time. So, instead of three satellites, the GPS receiver must receive four, so it can account for what's known as the receiver clock drift.

History

The GPS system was conceived in the 1970s but was not fully operational (a minimum constellation of 24 satellites) until 1995. Receivers were nearly five times as expensive to operate in the early days as they are now.

The cost of small-screen technology, such as light-emitting diode (LED) screens, came down with the proliferation of mobile phone technology in the 1990s, and with mapping data becoming cheaper or even free.

It was not until these technologies became cheaper that GPS receivers really became a consumer product.

Almost as soon as that happened, smartphones equipped with GPS came along and now most people are able to do 'turn-by-turn' navigation with an app in their phone.

Uses

GPS was originally designed to provide accurate position data to the US Army, Navy and Air Force, but has since been used for many applications it was never designed for.

It is used to avoid collisions in shipping, with all ocean-going vessels required to report via something known as the Automatic Identification System.

GPS is being used for improved landings of the most modern commercial aircraft. It has revolutionised fishing, allowing fishers to return to the exact same spot they'd fished before.

It is used in applications that allow people to keep track of others, their children, elderly relatives and pets, at a low cost.

GPS enables knowledge used for everyday convenience, such as when the next bus is coming, but also has exceptional capabilities for emergency operations such as for search and rescue.

By combining GPS with other technologies from robotics, self-driving cars are becoming cheap, and big mining companies are operating fully automated mines. For example, the West Angelas mine in the Pilbara uses automated trucks, rock-breakers, drills and blasters, all controlled from an operations centre in Perth.

A whole new category of industry has sprung up combining GPS with mobile communications: location-based services. This can mean finding the nearest toilet, hooking up with a blind date, keeping track of your running and cycling times or golf shots, or getting a review for the restaurants you're near.

Location-based gaming is also growing. This can range from chasing and hunting games to driving a simulated car on your computer against real drivers of big races in real-time.

Innovations in GPS

GPS is currently being upgraded for civilians, with two new signals which will make highly accurate position data available cheaply.

Russia's Global Navigation Satellite System became operational at almost the same time as GPS in the late 1990s but fell into decline due to economic problems in Russia. It is fully operational again now, and is adding new signals.

Europe (with Galileo) and China (with Beidou) are also working on similar systems. India and Japan are working on systems that are not 'global'. These systems are aimed at overcoming dependence on GPS (some 6–7 per cent of Europe's GDP – about €800 billion – relies in some way on satellite navigation). They should all be fully operational by 2020.

Concerns

With GPS being installed in phones, computers, cars and other high-value items, it is becoming much easier to track those items if they are stolen. The downside is that you may not want to be tracked. Location privacy is becoming an increasing issue.

An illegal remedy for privacy is the 'personal privacy device', which transmits a signal that 'jams' GPS.

Given some of the safety-critical applications mentioned above, it is clear that such devices – in the wrong hands – could present a threat to life. In one alarming US case, flight trials of a GPS-based aircraft landing system were interrupted several times by a jammer. It belonged to a courier who used it to avoid having his vehicle tracked by his employer. But its unintended consequence was to deny GPS precision guidance to the aircraft landing at a nearby (very large) airport.

Without doubt, the uses and potential abuses of GPS will progress in line with the technology.

What is the Human Genome Project?

Melissa Southey, *Professor of Pathology, University of Melbourne*

For many decades humans have pursued work to characterise the human genome – the complete set of human genetic information. Today, references to genome sequences are publicly available and have been instrumental in effecting recent advances in medicine, genetics and technology.

But interpretation of the human genome is in its early stages. Large initiatives are now embarking on more complex efforts to characterise the human genome, and understand individual genome variation.

What is the human genome sequence?

The human genome sequence is contained in our DNA and is made up of long chains of 'base pairs' – two chemical bases bonded to one another forming a 'rung of the DNA ladder' that makes up our 23 chromosomes. Along our chromosomes are the base pair sequences that form our 30 000 genes.

All humans share a great degree of similarity in their genome sequences – the same genes are ordered in the same manner across the same chromosomes, yet each of us is unique (except for identical twins) in terms of the exact base pair sequence that makes up our genes and thus our DNA/chromosomes.

It is this similarity that, in a genetic sense, defines us as human and the specific variation that defines us as individuals.

Launching the Human Genome Project

As early as the 1980s, momentum was gathering behind activities that supported, and would eventually define, the Human Genome Project.

Conversations had turned into workshops that likened characterisation of the human genome to the characterisation of the human anatomy that had centuries earlier revolutionised the practice of medicine.

In 1990, with continued support from the United States Department of Energy, the United States National Institutes of Health (NIH) and widespread international collaboration and cooperation, the $3 billion Human Genome Project was launched.

The project aimed to determine the sequence of the human genome within 15 years. By 2000 (well ahead of schedule) a working draft of the human genome was announced. This was followed by regular updates and refinements and today we all have access to a human 'reference genome sequence'.

This sequence does not represent the exact sequence of the base pairs in every human; it is the combined genome sequence of a few individuals and represents the broad architecture of all human genomes that scaffolds current and future work aiming to characterise individual sequence variation.

The detail and stories behind the Human Genome Project are themselves extraordinarily human. This project benefited from our human drive for discovery and advancement and our human response to competition.

It forced us as individuals and communities to consider our personal, ethical and social attitudes towards the availability of human genome information, intellectual property protection (especially gene patenting) and public versus private/commercial enterprise in a broad sense.

Advancing the project's success

In the years after the initiation of the Human Genome Project there were constant and significant advances in key areas that facilitated the enormous DNA sequencing effort.

These advances were achieved in all areas key to the efficient processing of DNA into electronic DNA sequence information. They included:

- improvements in the chemistries and instruments used to decipher the base pair sequences of prepared pieces of DNA;
- significant improvements in the capacity of computing facilities to manage the volume and nature of data generated from the instruments; and
- perhaps most importantly, improvements in the analytical tools.

The then state-of-the-art DNA sequencing chemistry used in the Human Genome Project was Sanger sequencing, which could sequence single stretches of several hundred base pairs at a time.

Advances in analytical methods of putting these pieces back together into the 3.3 billion base pair human genome were fundamental to the progress of the project.

The Human Genome Project was also advanced by competition. In 1998 a privately funded project with similar aims was launched in the United States by Craig Venter's Celera Genomics.

Using a modification of the DNA sequencing technique and a smaller budget, it was partly responsible for the accelerated progress of the Human Genome Project.

This competition brought forward other aspects of the project for ethical and legal scrutiny and discussion.

Patent wars

The issue of patenting genes formed a background to the Human Genome Project and many other similarly focused projects for some time. In the early 1990s it had been a serious issue of contention between James Watson (co-winner, with Francis Crick, of the Nobel Prize for discovering the structure of DNA) and Bernadine Healy (then Director of NIH).

Competition between Celera Genomics and the Human Genome Project now brought the discussion into a different dimension.

The publicly funded Human Genome Project released new data freely and in 2000 published the first working draft of the genome on the web.

In contrast, Celera filed preliminary patent applications on more than 6000 genes and also benefited from the data provided by the publicly funded project.

In March 2000, US president Bill Clinton announced that the genome could not be patented and should be made freely available.

The stock market dipped briefly because this announcement did not reflect the tangible benefits for biological research scientists.

Within 24 hours of the release of the first draft of the human genome, the scientific community downloaded half a trillion bytes of information from the University of California, Santa Cruz's genome server, which contains the reference sequence and working draft assemblies for a large collection of genomes. This was a strong indication of the relevance of this information to the biological, biotechnological and medical research communities.

Interpretation of the genome sequence is in its early stages but has already improved our ability to offer genetic testing and clinical management of many diseases.

We are now embarking on more complex pursuits to characterise the human genome so as to understand individual genome variation. This work is supported by projects related to, and of the same magnitude as, the Human Genome Project, including projects characterising the genomes of other species, among them mice and yeast and the International HapMap Project, which will describe the common patterns of human DNA sequence variation. Others include the Personal Genome Project, which works with volunteer subjects to make personal genome sequencing more affordable, accessible and useful; and the 1000 Genomes Project, which aims to find most genetic variants that have frequencies of at least 1 per cent in the populations studied.

These projects are greatly enhanced by the next generation of sequencing methodologies, which will expedite the characterisation of the human genome at an individual level in coming years.

Space, time and matter

'The fact that we live at the bottom of a deep gravity well, on the surface of a gas covered planet going around a nuclear fireball 90 million miles away and think this to be normal is obviously some indication of how skewed our perspective tends to be.'

Douglas Adams, *The Salmon of Doubt: Hitchhiking the Galaxy One Last Time*

'In the beginning there was nothing, which exploded.'

Terry Pratchett, *Lords and Ladies*

'Physics is like sex: sure, it may give some practical results, but that's not why we do it.'

Richard P. Feynman

'It is often stated that of all the theories proposed in this century, the silliest is quantum theory. In fact, some say that the *only thing that quantum theory has going for it is that it is unquestionably correct.'*

Michio Kaku, *Hyperspace: A Scientific Odyssey Through Parallel Universes, Time Warps, and the Tenth Dimension*

Black holes

Robert Braun, Chief Scientist for CASS, CSIRO

The concept of a 'black hole' is one of the most curious in astrophysics. It's the answer to the question: 'What happens if the density of matter in a region becomes so high that not even light can escape?'

The reason this question even arose dates back to Einstein's prediction of 1916, in his 'The Foundation of the Generalised Theory of Relativity', that the direction light travels will be bent in the direction of any nearby mass.

That prediction has been spectacularly confirmed in recent years by the discovery of 'gravitational lenses' – where a background source and a foreground mass are so closely aligned that the light from the background source is highly distorted, to the point of forming an almost complete arc around the foreground mass.

Before 1916, we had not considered that this might be possible. After all, why should photons, mass-less particles of light, feel the influence of any mass they pass? The surprising answer is that the basic shape of our universe is influenced by each mass within it, like the dimples in a rubber sheet caused by an occasional marble.

Passing light can be thought of as confined to the rubber sheet. When it passes near a marble it will be deflected from its original path by passing through the dimple. If the dimple becomes too deep, light that passes sufficiently close will be deflected into a spiral path that ends on the marble. It will not escape at all.

The boundary between light paths that just manage to pass a density concentration and those that do not is called the 'event horizon' – the greater the mass of a density concentration, the larger the size of this 'surface of no return'.

For an object as massive as our sun, which is about 1 391 980 km in diameter, the event horizon is about 6 km in diameter. The matter density – its mass divided by its volume – needed to form such a black hole is extremely high – about 2×10^{19} kg per cubic metre. That's more extreme than the density of an atomic nucleus.

The densest form of matter so far observed in nature is that encountered in so-called neutron stars, an entire star which has run out of fuel, collapsed and is now composed only of neutrons. Yet even a neutron star is not dense enough, by about a factor of 50, to form a black hole of the sun's mass.

Density

Curiously, the density needed to form a black hole scales as the inverse square of the total mass. So, the density of a neutron star would be sufficient to form a black hole if the object had about 2500 times the mass of the sun. What exactly happens when so much mass is concentrated in the same place is not understood.

Will it compress into some new state, such as the quarks that are thought to be the building blocks of neutrons, or some even more fundamental building block? We just don't know.

The average density of the sun – about 1 g per cubic centimetre – which also happens to be the density of liquid water on Earth's surface, would be sufficient to form a black hole were it associated with a mass of 100 million times that of the sun.

This is the mass of the compact objects within the centres of massive galaxies. Continuing this line of reasoning, we can ask: 'Do we live in a black hole?' – a question that's dominated cosmology for the past century.

Is the average density of the universe so high that light and everything else around us could never escape? The answer appears to be 'no'. The universe is undergoing an accelerating rate of expansion that implies an insufficient matter density to represent a black hole.

How do we know black holes exist?

What evidence is there that the black hole phenomenon actually occurs? At present, it's all indirect.

There are cases where a normal star appears to orbit a compact object that's very faint but is perhaps 10 times as massive as the sun. Since no current theories explain this, such objects have been called 'black hole candidates'.

On larger scales, there's evidence of a massive but compact object at the centre of many, and possibly all, galaxies. Sagittarius A, the object that resides in the very centre of our Milky Way, has been studied by tracking the movements of many nearby stars.

The orbits of these stars have been used to deduce that the central object must be about 4 million times more massive than the sun and that it must be smaller than about 1/1000 of a light year (a light year being equal to about 10 trillion kilometres).

Although this is the best current evidence for a very massive, very compact object, it's important to note that this limit on matter density, of about 10 g per cubic metre, is still 100 million times smaller than what's needed – about 1 kg per cubic centimetre – to qualify as a black hole of this mass.

Supermassive black holes

It's thought so-called supermassive black holes – as much as a billion times the mass of the sun – reside in the centre of galaxies that are significantly more massive than the Milky Way, particularly those which have an elliptical rather than a disk-like distribution of their stars. Many of these massive galaxies have been host to the 'quasar phenomenon' at some point in their history.

The quasar phenomenon is the most energetic type of event yet witnessed in the universe, outshining all of the stars in the hosting galaxies for millions of years, and is understood as a consequence of matter falling into a central massive object.

The extremely strong gravitational attraction of compact massive objects tends to tear apart and pull in anything that comes too close. The tearing action is due to 'tidal forces', the fact the gravitational force acting on the nearest portions of an object is significantly stronger than that acting on the most distant portions.

This same phenomenon causes tides on Earth, since the gravitational attraction of the moon is significantly larger on the side of Earth facing the moon than the side facing away. As material is pulled towards the compact massive object it tends to gather in what is called an 'accretion disk', a very hot, rapidly rotating structure that channels material towards the central object.

The reason for the rapid spin is the preservation of angular momentum, akin to what a slowly rotating ice skater experiences when they draw in their arms from a more extended position. Any small initial rotational motion is strongly amplified during contraction.

The high temperature is the result of the high-speed collisions between material falling in and what's already there.

The final stage of channelling material inward is challenging, since the energy associated with the rapid rotation must first be shed. The solution nature has found to this problem is dramatic: the rotational energy of the accretion disk is shed by sending high-speed jets of matter out along the rotation axis of the disk.

The same jet ejection phenomenon is found to apply over an extremely wide range of scales, from the formation of individual stars, such as the sun, to quasar accretion disks, where the result can be the largest distinct objects yet seen, measuring millions of light years from end to end.

But are the compact massive objects seen on a wide range of scales truly black holes, in the sense of having achieved a sufficient mass density to become disconnected from the rest of the universe? Or are they merely a highly condensed state of matter that we don't yet understand?

Seeing is believing

Direct evidence for the black hole phenomenon might be possible if an image could be made of the event horizon. While this has not yet been done, the best prospects might come from looking at the object at the centre of our own galaxy, since it provides the best combination of a large event horizon size with the closest possible distance.

The expected image size is about 0.2 milli-arcseconds (a 1.8 millionth of a degree, or about 10 000 times smaller than the typical image size of a star observed with an optical telescope from the ground.

The required resolution could be achieved by a network of radio telescopes separated by thousands of kilometres and observing at wavelengths of about a millimetre. Astronomers at the Smithsonian Institution are developing just such a project – the

Event Horizon Telescope (EHT). If successful, the EHT will demonstrate that nature has truly found a way to squeeze matter together to a density that even light cannot escape.

Probing the detailed shape of the event horizon and how it depends on total mass should provide clues about the state of matter under these extreme circumstances; circumstances that we cannot approximate in a laboratory.

Nature is likely to have some surprises waiting for us when we do. Each time we peel off a layer of the onion there seems to be another one inside.

Einstein's Theory of General Relativity

Jonathan Carroll, *Postdoctoral Research Associate, Centre for the Subatomic Structure of Matter, University of Adelaide*
Lewis Tunstall, *PhD Candidate, Centre for the Subatomic Structure of Matter, University of Adelaide*

It's 100 years from now. You wake up alone in a small, windowless room. The only other thing in the room is a small ball. Maybe the room is located in your city, but maybe it's inside that new spaceship everyone's talking about. How can you tell?

You pick up the ball and drop it. It falls vertically to your feet. You time the fall and calculate that the ball accelerates at 9.8 metres per second per second (9.8 m/s^2), exactly the acceleration of gravity at the surface of the earth.

But a spaceship in the middle of deep space can also accelerate by that much, producing the exact same results. So where are you?

In 1911, Einstein formally proposed that gravitational mass (the kind that produces a gravitational field) and inertial mass (the kind that resists acceleration) were one and the same. This became known as the 'equivalence principle'. According to this principle, you can't tell whether you're in a gravitational field (such as on the surface of the earth) or experiencing constant acceleration (a spaceship speeding up, pushing you to the floor, like the g-force of a roller-coaster).

Another example is the infamous 'Vomit Comet' (officially the 'Weightless Wonder'), used by NASA for training, and occasionally by Hollywood for filming. Just as in our example of the ball, there's no way to tell the difference between free fall, and being in the absence of a gravitational field, say, in deep space.

This principle led Einstein to consider incorporating gravity into the framework of his Special Theory of Relativity, culminating in his Theory of General Relativity.

At face value, that doesn't appear such a difficult thing to do. Until this point, the properties of objects in isolation could be described by equations with great accuracy. But what to do about gravity? How does one calculate the properties of a system in which acceleration can be due to either gravity or changes in velocity? It seems to depend on how you are looking at it.

That led to the idea of a 'reference frame' – the stage on which the objects you are looking at play out their roles. There may of course be other frames in which the objects appear to behave differently, so we need a description of all the frames, and the way to relate them.

The trick was to consider space and time as a four-dimensional object in itself – not a fixed stage on which the objects are defined, but something that itself can change.

Space–time

Let's say you and I are going to meet for coffee. How do you describe this 'event'? One option is to look at a map – 'I'll meet you at the cafe on level two of the building that's at G5 on the map.' We have described three coordinates: G, 5, and level two. This is another way of saying a set of x, y, and z coordinates. So that we actually meet for coffee, we'll also need to add a fourth coordinate: time – say 2:00 pm. These four points are what we call a space–time event.

General Relativity says the map can be distorted; and our coordinates will depend on how that happens. If I were to bend the map a little, the distance between two locations changes.

If you measure and add the angles of a triangle on the flat map you would get 180 degrees. If you do this on the curved map, you get a little more or a little less (depending on which way it's curved). In the same way, the universe itself can have areas of different curvature.

Now for the mind-bending part …

You might know that, in the absence of any external forces, things like to travel in straight lines. But what about when the space is curved? We can still talk about straight lines, but now the lines follow the curvature. Think about drawing a small, straight line on a basketball. You can draw a line all the way around the ball and arrive back at the starting point. It's straight, but also curved.

Odd things happen in 'curved space' that contradict what we expect from 'flat space'. If you walk north 10 kilometres, west 10 kilometres then south 10 kilometres, you would expect to end up 10 kilometres west of where you started. Do that at the South Pole and you end up where you started! Technically this happens everywhere, but on a (non-cylindrical projection) map it's obvious at the poles.

Now we can expand our definition and say objects not influenced by a force travel along straight lines in curved space. In particular, things with mass (or energy, thanks to $E = mc^2$) follow these straight paths in curved space.

The experimental proof of this occurred during a solar eclipse in 1919 where starlight was observed to be bent by the sun. Einstein predicted the amount of bending, and not by the standard 'Newtonian' theory.

So matter follows the curvature of space, but we know matter is the source of gravity, so the curvature responds to matter as well. In the words of American theoretical physicist John Archibald Wheeler, 'Matter tells space–time how to curve, and curved space tells matter how to move'.

What if we have lots of matter in one place? Imagine you are driving up a steep hill. Some steepness is too much for your car to manage, even at its fastest. In the same way, if we have a very large amount of matter in a very small area, the curvature becomes so strong that not even light is fast enough to get out. This is a black hole.

Beautiful curves

Starlight and black holes are fun, but what does this have to do with day-to-day life on Earth? Have you ever used the Global

Positioning System (GPS)? It's a common feature of mobile phones today, and it relies entirely on General Relativity to work.

We said our map could be curved so that the points in the space dimensions were closer together. Since space and time behave together as space–time, the same trick happens for time. If we have some large mass, the curvature in the time dimension means that the more curved the space–time is, the slower a clock ticks there (or appears to for someone in a less curved region).

There is a measurable difference between the rate at which your atomic clock ticks on the surface of the earth, and that at which one in orbit ticks.

Without this correction, GPS satellites would not be able to tell you where you are with such accuracy.

General Relativity has seen so many experimental achievements with astounding precision (explaining the anomalous orbit of Mercury, orbital decay of binary stars, and the gravitational redshift of light) that it's hard to believe it might not be the complete theory of gravity.

Some speculation arose when NASA's Pioneer 10 and 11 spacecraft (at around 15 400 000 000 and 12 400 000 000 km from Earth, respectively) appeared to be slowing down almost imperceptibly – more so than would be expected, even taking into account General Relativity effects.

But it appears that thermal radiation from the crafts is slowing them slightly, and General Relativity remains intact.

General Relativity is possibly one of the most comprehensive theories ever formulated, and certainly involves many more facets than can be covered here. Gravity waves, gravitational lensing, dark energy (and the fact it cannot be combined with the standard model of particle physics) are all thoroughly interesting topics.

But to have the time to describe them all, we would need to be accelerating near the speed of light … or in a strong gravitational field.

Gravity

David Blair, Director, Australian International Gravitational Research Centre, University of WA

I have spent almost 40 years trying to detect gravity waves.

When I started there were just a few of us working away in university laboratories. Today 1000 physicists working with billion-dollar observatories are quietly confident the waves, predicted by Einstein in 1916, are within our grasp.

If we are right, the gravity wave search will have taken 100 years from the date of Einstein's prediction.

In 100 years' time the discovery of Einstein's gravity waves will be one of the landmarks in the history of science. It will stand out like the discovery of electromagnetic waves in 1886, a quarter of a century after physicist James Clerk Maxwell predicted them.

The problem with gravity waves is that you can't explain them without explaining Einstein's idea of gravity. Recently I began to ask why it is so difficult to explain gravity, why the concept is met with glazed eyes and baffled looks. Eventually I came up with a theory I call the Tragedy of the Euclidean Time Warp.

Discarding Euclidean ideas

My theory starts 2300 years ago with Greek mathematician Euclid's book of geometry, *Elements* – the most influential book in the history of science.

Elements has been in print for more than 2000 years and published in more than 1000 editions. It was a basic school text for Galileo, Newton, Einstein and every educated person up to the baby-boomer generation. I still have the plain, slim edition I used in year eight.

The basic concepts from *Elements* are still taught in all primary schools and high schools throughout the world every day. We all know those concepts – parallel lines never meet, the sum of the angles of a triangle is 180°, the Theorem of Pythagoras and the perimeter formula for a circle ($P = 2\pi r$).

Euclidean geometry has moulded the way we think. Today we all have a conception of space defined by Euclidean geometry.

The problem with Euclid's book is that it cements a false idea about space. It is a shock to think that it is wrong but even so, gravity cannot be explained without discarding Euclidean geometry.

The link to General Relativity

The possibility of a flaw in Euclid's book was first raised by mathematician Carl Gauss in the 1820s. He published a theorem that said you could measure the shape of space by measuring angles and distances. He even tried to measure the shape of space on Earth by measuring the sum of the angles of a triangle between three mountain-tops.

Some 90 years later, Einstein published his Theory of General Relativity, which gave us our current explanation for gravity. His theory is conceptually simple, but mathematically complex. Matter curves space and time, and gravity arises because of the way matter floats in this deformed space.

One of the key observations that confirmed the theory of curved space was made in Australia in 1922. The Wallal expedition obtained photos during a solar eclipse, from which the bending of the light as it passed by the sun was measured.

I think is rather shocking that today, 90 years later, we still teach geometry as if space were flat.

The reason our culture has not assimilated curved space is that we were all indoctrinated with Euclidean geometry during childhood. By the time we are adults it takes a painful re-think to adapt to a new way of thinking.

Trainee school teachers are rarely exposed to General Relativity, so the teaching profession remains entirely free from Einstein's beautiful theory. Generation after generation, this cycle continues. We are trapped in this Euclidean time warp. This is a tragedy not only because truth is important, but because students are disengaged by the stale teaching of obsolete 19th century physics.

A primary school experiment

I've tried to catch 11-year-olds before they were indoctrinated. Rosalie Primary School, in suburban Perth, agreed to host me, and I began weekly sessions with 30 primary school kids for six weeks. Here is how we learnt that the force we call gravity arises because time is warped by matter.

First we talked about straight lines. How do we tell if lines are straight? Can you draw straight lines on balloons or the surface of the earth? What do surveyors do when they are building a straight fence? They always use sight lines, and when it comes down to it, straightness is always measured with light.

Then we considered drawing triangles on the earth. Suppose you start at the North Pole, travel south to the equator, turn left and travel 90° of longitude eastwards before taking another right turn to head back to the pole. We could all see that this triangle would have 90° + 90° + 90° – a total of 270° and definitely not what Euclid said.

Einstein said we should think about space and time together – what we call space–time. But space–time has four dimensions and we all agreed our brains just don't work properly for four dimensions.

So instead we agreed we could use just distance and time to keep it simple. We could then draw the space–time diagram for the journey to school or for a water balloon falling from the Leaning Tower of Gingin, a full size steel replica of the Leaning Tower of Pisa, at an upcoming excursion.

Everyone could draw a space–time diagram for their journey to school and point to the places where distance was not increasing but time was passing – waiting at the lights – and the steep bits when they were speeding down the freeway.

The next step in explaining Einstein's theory of gravity is to think about the length of a trajectory in space–time. Einstein said things in free-fall always have the shortest trajectory in space–time. At first sight this is a weird idea. How can you measure a trajectory when one axis is distance and the other axis is time?

The journey-to-school diagram looks completely different if you change your units from seconds to minutes or metres to miles, and the idea of the length of trajectory is pretty meaningless when both axes have different units.

A Journey to School in Space–time

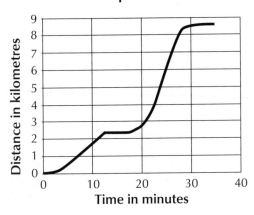

Science Education
Enrichment project

The only sensible way to measure space–time is to use a speed that enables us to measure time in space units. Using lightspeed as the conversion factor we can convert any time to the number of metres travelled by light in that time.

If you drop a water balloon from the Gingin Tower it takes almost three seconds to hit the ground. A space–time diagram of its trajectory is a parabola that starts 45 m above the ground, and hits the x-axis (time axis) three seconds later. But three seconds in time is $3 \times 300\,000$ km, or 9 million metres.

The balloon travelled 900 million metres in time. To full-scale the graph would stretch twice as far as the moon. The space–time diagram is extremely elongated!

So now we can plot space–time trajectories using metres for both distance and time, and I can now imagine using a tape measure to measure the length of any trajectory in metres.

Floating and falling in space

Now we come back to Einstein's theory. It says that if you allow something to float freely in space, or fall from a tower, its trajectory in space–time will always be the shortest.

Since a shortest line normally means a straight line, Einstein is saying that free-floating trajectories are 'straight lines' in space–time (technically they are called geodesics).

Anything you do to prevent free-floating (like holding on to the water balloon on top of the tower rather than releasing it) will make the trajectory longer.

Now comes the Eureka moment. For the water balloon on the tower I want you to retort indignantly: 'This is nonsense. It is obvious that the not-falling space–time trajectory is always shorter than the falling trajectory!'

In one case the balloon fell down 45 m, while in the other it did not even move in space, although it kept on going in time just as before. One trajectory was roughly diagonal on the space–time graph but the other just moved parallel to the x-axis. This is like saying the hypotenuse of a right-angled triangle is shorter than its sides.

I reply: it is not nonsense! Einstein is correct because time depends on height above the earth. Time is warped. Time on the top of the tower is running four parts in one million billion times faster than it is on the ground.

Another way of saying this is that a clock on the top of the tower runs faster by four femtoseconds (4×10^{-15} seconds) for every second. This is enough to stretch the time axis by just enough that not falling is actually longer than falling.

And there is yet another way of saying this. Gravity is the force you have to apply to objects to prevent them from freely floating in space. You do not feel gravity while you are falling because there is no force. You are just following the shortest path in space–time.

What we call gravity is the result of the warping of time by the mass of the earth. Gravity is a force exerted by the earth to stop you from falling.

We all have a destination in space–time called old age. Nature tries to make things get to this destination as quickly as possible.

An astronaut becomes an old astronaut more quickly if he is floating around in the space station. The astronaut is higher above the surface of the earth than the top of the Gingin Tower. Time on the space station runs even faster than it does on top of the tower. Therefore the astronaut ages more quickly there than on the earth's surface or at the top of the tower. It takes forces to delay the progress of time.

We have a good planet that reliably provides continuous forces to us to prevent us free floating (i.e. falling) to the centre of the earth. So our planet is a time machine that delays our ageing (but just by a millisecond in a lifetime!).

The most astonishing thing about my program with Rosalie Primary School was that the kids were not astonished. My colleagues Grady Venville and Marina Pitts and I measured their learning and asked them if they thought they were too young to learn this stuff. By a large majority they thought they were not too young and that it was really interesting.

Physicists and astronomers deal with curved space every day, and even our GPS navigators have to correct for the warped space–time around the earth. In May 2011, NASA's Gravity Probe B spacecraft found that the perimeter of an orbit around the earth failed to match its Euclidean value by 28 mm – not a large discrepancy but just the value predicted by Einstein.

Not too difficult to teach

In spite of modern science, the general belief among educators is that Einstein's physics is too difficult to teach in school. As a result, science students enter university indoctrinated with 2300-year-old Euclidean geometry and 300-year-old Newtonian physics.

Very few go on to discover the Einsteinian reality of curved space and warped time. The lucky few who get to study Einsteinian physics have difficulties because the fundamental concepts contradict all their past learning.

Most students who go on to become teachers maintain the Newtonian mindset and so education remains in a Euclidean time warp! The drastic decline in science at school and university could in part be due to our failure to challenge young people with modern ideas such as these.

If we start young enough, everyone can easily learn that the world is non-Euclidean, and then appreciate that the geometrical formulae we learn at school, such as Newton's Law of Gravitation, are convenient approximations for everyday life.

Heisenberg's Uncertainty Principle

Howard Wiseman, *Professor in Physics, Griffith University*

The term 'uncertainty principle' suggests some grand philosophical idea, like 'you can never be sure of anything', or 'there are some things you can never be sure of', and sometimes people use it as if this is what is meant.

In fact, this principle, discovered by German theoretical physicist Werner Heisenberg in 1927, has a precise technical meaning that's typically relevant only to microscopic particles. But it also has implications for how we understand the universe and our relation to it, and also to new technologies of the 21st century.

Uncertainty about what?

While the Heisenberg Uncertainty Principle (HUP) does not mean 'there are some things you can never be sure of', it *does* imply 'you can never be sure of everything'. How can this be? If you can never be sure of everything, doesn't that mean there are some things you can never be sure of? Surprisingly, no.

In science we are ultimately concerned with what we observe. So when we say we are uncertain about something, we mean that we are uncertain about what we will observe when we do an experiment.

Of course, life would be pretty boring if we could always predict what was going to happen next, but for many centuries scientists dreamt they would be able to do this. The HUP killed that dream in a very interesting way.

The simplest example of the HUP is this: You can never be certain of both the position and the speed of a microscopic particle. It is possible to arrange an experiment so you can predict the position of a particle. A different experiment would let you predict its speed. But you will never be able to arrange things so that you can be certain of both its position and its speed.

You might be jumping up and down at this point, saying, 'That's ridiculous. If I want to know both I just measure them simultaneously. Or I first measure the position, then the speed.' In fact, neither of these options will work. What rules them out is other forms of the HUP.

In the first case, there is the HUP that says it is not possible to simultaneously measure position and speed with perfect accuracy. In the second, there is the HUP that says if you accurately measure the position you will disturb its speed, making it more uncertain, and vice versa. So you can't get around it.

Is it a principle?

Before getting into the details, one thing to get clear is that Heisenberg's 'Uncertainty Principle' is not really a principle at all. A 'principle', in science as in everyday life, is a fundamental simple idea from which all sorts of other things can be derived, such as the principle of freedom, or the principle of fairness.

Heisenberg's principle is not like that – it's actually a consequence of something more fundamental. That thing is quantum mechanics, a theory that applies to all forms of matter and energy (as far as we can tell).

Unfortunately, although quantum mechanics seems fundamental, it's not simple, and so cannot be encapsulated as a principle. But from it follow all forms of the HUP.

Precisely uncertain

For the example given earlier, Heisenberg's principle can be precisely stated as:

$$\Delta q \times \Delta v > \hbar/m \qquad (1)$$

Here Δq is the uncertainty in the position of the particle (in metres), Δv is the uncertainty in its speed (in metres per second), m is its mass in kg, and $\hbar$ is a constant (Planck's constant, which also turns up in the formula linking the amount of energy a photon carries with the frequency of its electromagnetic wave).

Note that the two uncertainties are multiplied together in Equation (1), and the result must be greater than some number. This means that, although Δq can be as small as you like as long as Δv is large enough, or vice versa, they cannot both be arbitrarily small.

The HUP in its looser form ('you can never be sure of everything') is thus a consequence of the fact that Planck's constant is not zero. But Planck's constant is very small. In the units used here, $\hbar \approx 10^{-34}$; that is 0.00 ... 001, where there should be 34 zeros here. This smallness is why we don't have to worry about the HUP in everyday life.

You may have heard the anecdote about a woman who is stopped by a policeman who says: 'I just measured your speed as 53.9 km h^{-1} in a 40 km h^{-1} school zone'. She retorts: 'Are you familiar with the Heisenberg Uncertainty Principle? If you are so sure about my speed, you can't possibly know where my car was.'

It's a cute joke, but let's see what the HUP actually says. When the policeman says the speed was measured as 53.9 km h^{-1}, he presumably just means it was closer to 53.9 than to 53.8 or 54.0. This means an uncertainty of about 0.05 km h^{-1}, which is about 0.01 m/second. If the mass of the car is 1000 kg then the HUP implies:

$$\Delta q > \hbar/(m \times \Delta v) \approx 10^{-34}/(1000 \times 0.01) = 10^{-35} \text{ metres}$$

Thus the minimum uncertainty in the position of the car implied by the HUP is much, much smaller than the size of an atom. So this is obviously irrelevant when it comes to the question of whether the car was in the school zone or not.

Although the HUP doesn't have much to say about speeding tickets, it's ubiquitous at the scale of atoms and subatomic particles. The mass of an electron is extremely small ($m \approx 10^{-30}$ kg) so that $\hbar/m \approx 10^{-4}$ on the right-hand-side of Equation (1) is no longer ridiculously small.

In fact, some simple arguments involving the motion of electrons around the nucleus of an atom let us derive the approximate size of an atom as the minimum $\Delta q \approx 10^{-10}$ metres implied by the HUP. The HUP in one form or another is a useful principle in almost every field of science dealing with very small amounts of matter or energy.

Applications in technology; implications in philosophy

Since quantum mechanics underlies almost all modern technology, the HUP turns up all over the place. It also plays a more direct role in the quantum technologies of the 21st century, which are just being developed now.

Quantum communication allows the sending of encoded messages that can't be hacked by any computer. This is possible because the messages are carried by tiny particles of light called photons. If an eavesdropper attempts to read out the message in transit, they will be discovered by the disturbance their measurement causes to the particles as an inevitable consequence of the HUP.

The HUP also raises fascinating and difficult philosophical questions. The most obvious is: What's the reason for this uncertainty?

In everyday life we could be uncertain whether the cue ball will end up going into the top pocket because we are uncertain about its speed or position. But we would not doubt that the ball has a speed and position.

In the regime of quantum experiments, by contrast, we are uncertain about the results of experiments because the particle itself is uncertain. It has no position or speed until we measure it. Or so Heisenberg thought, and most physicists still follow this line.

However, others strongly disagree with this conclusion. The debate is not over – that's for certain.

Quarks

Takashi Kubota, *Research Fellow in Experimental Particle Physics, University of Melbourne*

One of humanity's eternal questions surrounds what we are fundamentally made of. Many ancient philosophies believed in a set of classical elements: from water, air, fire and earth of ancient Greeks; to water, fire, earth, metal and wood of East Asian Wu-Xing thought.

Physicists today believe matter is made up of 12 fundamental particles – quarks and leptons (a class of particles that don't undergo strong interactions) – that have no substructure and cannot be broken down into smaller particles. Quarks and leptons interact via four forces to make the universe we know today.

How these particles work to make matter

Six types of quarks have so far been experimentally confirmed, and given the names 'up', 'down', 'strange', 'charm', 'bottom' and 'top', in ascending order of mass.

There also are six types of leptons, three electrically charged ones: 'electron', 'muon' and 'tauon'; and three electrically neutral ones, all called neutrinos. Each of the three neutrinos pairs up with one of the charged leptons and is then called 'electron neutrino', 'muon neutrino' and 'tau neutrino', respectively.

These twelve elementary particles are mediated by exchanges of another type of particle referred to as a force mediator. These can be described as the force between the particles. They can change the direction of motion of a particle, its mass and its electrical or other charge. Today, four types of elementary force mediators are known, namely the 'gluon', 'photon', 'graviton' and 'weak bosons'.

Actually, the graviton has not yet been experimentally confirmed – possibly because it interacts very weakly with mass – but many physicists assume this mediator exists.

In an atomic nucleus, a proton is made up of two up quarks and one down quark, and a neutron is composed of one up quark and two down quarks. The force that binds three quarks in a proton or a neutron is called the strong force, and is due to exchanges of gluons.

An atomic nucleus, together with the electrons orbiting around it, constitutes an atom. The relation between the nucleus and electrons resembles the one between the sun and planets in the solar system.

The nucleus and the electrons are attracted to each other, exchanging photons. The force between the nucleus and electrons is the electromagnetic force.

Many atoms constitute objects in our everyday life as well as much bigger components of the universe such as stars and galaxies. The force dominating this level of macroscopic phenomena is gravity, intermediated by gravitons.

In the centre of stars, huge energy is generated by nuclear fusion being mediated by weak bosons. This energy makes the universe bright. In nuclear fusion, a down quark is changed to an up quark by the weak force. Stars are luminous because the fundamental building blocks are changing their types and providing energy.

Quarks like to hang in groups

Although most physicists believe that quarks are the fundamental building blocks which make up the universe, no one has observed an isolated quark on its own. This is because of the nature of the strong force.

Like a nucleus and an electron that attract each other due to their electrical charges, quarks are combined together by their colour charges. The strong force is a force that works between colour charges. Just as there are two types of electrical charges, there are three types of colour charges – 'red', 'blue' and 'green' – analogous to the primary colours of light. The strong force forces quarks to be in a 'white' state.

Colour charge actually has nothing to do with visible colours, it is simply a convenient naming convention for a mathematical system physicists developed to explain their observations about quarks in hadrons.

If you have learnt the theory of the elementary colours of light, you will remember that the superposition of the three elementary colours ends up with white. This is the reason a proton and a neutron consist of three quarks. In a proton and a neutron, one quark has a red colour, another has a blue colour and the third one has a green colour.

As a consequence of the fact that the strong force prohibits non-white states, no one has succeeded in isolating a quark. This phenomenon is called the quark confinement.

The existence of quarks has been established by several experiments, but should you find a way to isolate an individual quark, you would be in line for a Nobel Prize.

Cutting edge

Some physics theorists seriously think about the possibility that in the primordial universe (around 10^{-6} seconds after the Big Bang), another phase was realised in which quarks and gluons were flying as free particles. This phase is called the quark-gluon plasma (QGP) phase.

Scientists have been trying to reproduce the QGP phase by colliding heavy ions using powerful particle accelerators like the Large Hadron Collider (LHC) in Geneva and the Relativistic Heavy Ion Collider (RHIC) in New York.

So far, the RHIC has indicated that the creation of a QGP phase is possible at the hottest temperature ever reached in a laboratory (it is 4 trillion degrees celsius, 250 000 times hotter than the centre of the sun).

Though quarks are the elementary particles in physics today, there have been trials to see if there is structure within them. If one observes the structure of a quark, it means that quark is no longer elementary but a composite particle consisting of more fundamental particles.

One way to see the structure would be to follow the manner of an experiment Sir Ernest Rutherford performed 100 years ago. He shot alpha particles at a gold foil and observed that some of them were deflected at a very large angle. This showed the existence of a hard core within an atom.

Likewise, an unexpected large deflection of incident particles in high-energy collisions might mean the discovery of the structure of quarks. The current most powerful particle accelerator is the LHC, which has been colliding protons at the centre of mass energy of eight teraelectron volts (one million times higher than the energy of alpha particles).

At present the LHC is shut down for an upgrade, but it will be back with its energy doubled in 2015. So stay tuned!

String theory

Dean Rickles, *Associate Professor, University of Sydney*

String theory entered the public arena in 1988 when a BBC radio series *Desperately Seeking Superstrings* was broadcast.

Thanks to good marketing and its inherently curious name and features, it's now part of popular discourse, mentioned in TV's *The Big Bang Theory*, Woody Allen stories, and countless science documentaries.

But what is string theory and why does it find itself shrouded in controversy?

Life, the universe and the theory of everything

Today we think of string theory in two ways.

It's seen as a theory of everything – that is, a theory that aims to describe all four forces of nature within a single theoretical scheme.

These forces are the:

- electromagnetic force;
- gravitational force;
- weak nuclear force; and
- strong nuclear force.

Electromagnetism and gravity are familiar to most people. The nuclear forces occur at a subatomic level, and are unobservable by the naked eye.

String theory is also used to describe quantum gravity, a theory that combines Einstein's theory of gravity and the principles of quantum theory.

Tangled beginnings

But string theory began life more modestly, as a way to describe strongly interacting particles called hadrons.

Hadrons are now understood to be composed of quarks connected by gluons but string theory viewed them as quarks connected by strings (tubes of energy).

Understood this way, string theory buckled under both new experimental evidence (leading to the crowning of quantum chromodynamics – so called because it labels three different kinds of charges as 'red', 'green' and 'blue', despite their having no actual colour – which describes the interactions of quarks and gluons) and also internal problems.

String theory involved too many particles, including a massless particle with spin 2 – spin being the name used for the quantum mechanical version of angular momentum.

As it happens, this is exactly the property possessed by the graviton – the carrier of gravitational force in the particle physics picture of the world.

Beyond four dimensions

This discovery meant that with a bit of skilful rebranding (and rescaling the energy of the strings to match the strength of gravitation), string theory shed its hadronic past and was reborn as a quantum theory of gravity.

All the other particles that were problematic for the original string theory were able to capture the remaining non-gravitational forces too. This is how string theory took on its current role as describing all four forces together: a theory of everything.

But it could not shed many of its curious features.

One such feature was that it postulated many more space–time dimensions than are actually observed.

In a 'bosonic' version of string theory (i.e. *without* matter or fermions – quarks, leptons, and any composite particle made of an odd number of these), there would have to be 21 dimensions – 20 space dimensions and one time dimension.

In a theory *with* fermions, there would have to be nine spatial dimensions and one temporal, 10 dimensions altogether.

The problem is that we only perceive four dimensions: height, width, depth (all spatial) and time (temporal).

Supersizing symmetry, downsizing dimensions

The 'super' in 'superstring theory' refers to a kind of symmetry, known as supersymmetry, which relates bosons and fermions, suggesting that every particle has a 'superpartner' particle with slightly different characteristics.

There are five possible theories that involve matter in 10 dimensions. This was previously seen as a problem since it was expected that a theory of everything should be unique.

The six unseen dimensions (10 minus the four dimensions of everyday life) are made too small to be observable: a process known as compactification.

Beautiful maths

It is from this process that much of the extraordinarily beautiful (and fiendishly difficult) mathematics involved in string theory stems.

We have no trouble thinking of each event in the world as labelled by four numbers or coordinates (e.g. x, y, z, t).

A string-theoretic world adds another six coordinates, only they are crumpled up into a tiny space of radius comparable to the string length, so we don't see them.

But, according to string theory, their effects can be seen indirectly by the way strings moving through space–time will wrap around those crumpled, curled-up directions.

There are very many ways of hiding those six dimensions, yielding more possible stringy worlds (perhaps as many as 10^{500}!).

How long is a piece of string?

This is why string theory is so controversial. It seemingly loses all predictive power since we have no way of isolating our world amongst this plenitude.

And what good is a scientific theory if it cannot make predictions?

One response is to say that these various theories are not really so different. In fact there are all sorts of exact relations known as dualities connecting them.

More recent developments based on these dualities include a new type of object with higher dimensions – so called Dp-branes, which open strings can end on.

These too can wrap around the compact dimensions to generate potentially observable effects.

Most importantly, they can also provide boundaries on which endpoints of strings sit.

Just to complicate things more, a new kind of theory has been discovered, this time in 11 dimensions: 11-dimensional supergravity – it is also very beautiful mathematically.

Dial M for multiverse

String theorists are fond of saying that these six theories are aspects (special limits) of a deeper underlying theory, known as M-theory. In this way, uniqueness is restored.

Or is it?

We still have the spectre of the 10^{500} solutions or worlds. The great hope is that the number of solutions with features like our own world's (with its four visible dimensions, particles of various types interacting with particular strengths, conscious observers, and so on) will be small enough to be able to extract testable predictions.

So far, though, the only real way of getting our world out of the theory involves the use of a multiverse (a realistically interpreted ensemble of string theoretic worlds with differing physical properties) combined with the anthropic principle – which broadly states that the universe must have those properties that allow life to develop in it at some stage in its history. But only some of these worlds have what it takes to support humans.

Needless to say, this does not entirely sit easily with critics of string theory!

But string theory has been making strides in other areas of physics, notably in the physics of plasmas and of superconductors.

Whether this success can be repeated within its proper realm of fundamental physics remains to be seen.

The Doppler effect

Gillian Isoardi, *Lecturer in Optical Physics Science and Engineering Faculty, Queensland University of Technology*

When an ambulance passes with its siren blaring, you hear the pitch of the siren change: as it approaches, the siren's pitch sounds higher than when it is moving away from you. This change is a common physical demonstration of the Doppler effect.

The Doppler effect describes the change in the observed frequency of a wave when there is relative motion between the wave source and the observer. It was first proposed in 1842 by Austrian mathematician and physicist Christian Johann Doppler. While observing distant stars, Doppler described how the colour of starlight changed with the movement of the star.

To explain why the Doppler effect occurs, we need to start with a few basic features of wave motion. Waves come in a variety of forms: ripples on the surface of a pond, sounds (as with the siren above), light and earthquake tremors all exhibit periodic wave motion.

Two of the common characteristics used to describe all types of wave motion are wavelength and frequency. If you consider the wave to have peaks and troughs, the wavelength is the distance between consecutive peaks and the frequency is the count of the number of peaks that pass a reference point in a given time period.

When we need to think about how waves travel in two- or three-dimensional space we use the term wavefront to describe the linking of all the common points of the wave.

So linking all the wave peaks that come from the point where a pebble is dropped in a pond would create a series of circular wavefronts (ripples) when viewed from above.

Consider a stationary source that's emitting waves in all directions with a constant frequency. The shape of the wavefronts coming from the source is described by a series of concentric, evenly spaced 'shells'. Any person standing still near the source will encounter each wavefront with the same frequency it was emitted at.

But if the wave source moves, the pattern of wavefronts will look different. In the time between one wave peak being emitted and the next, the source will have moved so that the shells will no longer be concentric. The wavefronts will bunch up (get closer together) in front of the source as it travels and will be spaced out (further apart) behind it.

Now, a person standing still in front of the moving source will observe a higher frequency than before as the source travels towards them. Conversely, someone behind the source will observe a lower frequency of wave peaks as the source travels away from it.

This shows how the motion of a source affects the frequency experienced by a stationary observer. A similar change in observed frequency occurs if the source is still and the observer is moving towards or away from it.

In fact, any relative motion between the two will cause a Doppler shift/effect in the frequency observed.

So why do we hear a change in pitch for passing sirens? The pitch we hear depends on the frequency of the sound wave. A high frequency corresponds to a high pitch. So while the siren produces waves of constant frequency, as it approaches us the observed frequency increases and our ear hears a higher pitch.

After it has passed us and is moving away, the observed frequency and pitch drop. The true pitch of the siren is somewhere between the pitch we hear as it approaches us, and the pitch we hear as it speeds away.

For light waves, the frequency determines the colour we see. The highest frequencies of light are at the blue end of the visible spectrum; the lowest frequencies appear at the red end.

If stars and galaxies are travelling away from us, the apparent frequency of the light they emit decreases and their colour will move towards the red end of the spectrum. This is known as red-shifting.

A star travelling towards us will appear blue-shifted (higher frequency). This phenomenon was what first led Christian Doppler to document his eponymous effect, and ultimately allowed Edwin Hubble to propose that the universe is expanding, after he observed that all galaxies appeared to be red-shifted (i.e. moving away from us and each other).

The Doppler effect has many other interesting applications beyond sound effects and astronomy. A Doppler radar uses reflected microwaves to determine the speed of distant moving objects. It does this by sending out waves with a particular frequency, and then analysing the reflected wave for frequency changes.

It is applied in weather observation to characterise cloud movement and weather patterns, and has other applications in aviation and radiology. It's even used in police speed detectors, which are essentially small Doppler radar units.

Medical imaging also makes use of the Doppler effect to monitor blood flow through vessels in the body. Doppler ultrasound uses high frequency sound waves and lets us measure the speed and direction of blood flow to provide information on blood clots, blocked arteries and cardiac function in adults and developing foetuses.

Our understanding of the Doppler effect has allowed us to learn more about the universe we are part of, measure the world around us and look inside our own bodies. Future development of this knowledge – including how to reverse the Doppler effect – could lead to technology once only read about in science-fiction novels, such as invisibility cloaks.

The fifth dimension

Samuel Baron, *Postdoctoral Fellow, University of Sydney*

By now we're used to the idea that the world has four dimensions: three spatial and one temporal. But what if there were a fifth dimension – what would that dimension look like, and how would it relate to time?

One of the central threads running through the philosophy of time concerns the idea that time flows. This doesn't sound controversial: for most people, it flows in much the same way as a river.

But there are problems with this view.

If time flows, it's surely reasonable to wonder about the rate it flows at.

But rates of flow are construed as ratios over time – a river flows at one metre per unit of time, say – so it would seem time should also flow at some rate over time: one second per second.

This doesn't work, though. It's like saying that, for every dollar you give me, I give you a dollar back: what could be gained in such a transaction?

We commonly think of time flowing into the future, away from the past, but time wouldn't 'go anywhere if it flowed at one second per second.

What to do?

One option might be to construe the flow of time as a ratio of time over space, so that time flows, for example, at one second per metre.

But this too would be pretty odd: treating time as dependent on space in this fashion flies in the face of our intuitive understanding of its nature.

Another alternative might be to invoke a further temporal dimension – a fifth dimension – which can then be used as the yardstick for measuring temporal flow.

Call the ordinary temporal dimension 'A' and this new temporal dimension 'B'. In this view, time flows at one second of A per second of B. Thus, time now 'goes somewhere' in that it charts a path through a higher dimension.

Appealing to a fifth dimension in this fashion is often seen as a strategy of last resort by philosophers of time: the idea is simply too wild to take seriously.

The LHC

It has been predicted that the Large Hadron Collider in Switzerland might generate particles that time-travel by taking shortcuts through a fifth dimension.

Spatial treatment

Unfortunately, according to the physicists responsible for the relevant experimental predictions, the fifth dimension is spatial, not temporal.

Even if the fifth dimension *were* temporal, there'd still be a problem. This is because the experimental predictions conducted so far are produced against the backdrop of a particular physical theory: General Relativity.

General Relativity undermines any basis we might have for believing that time flows at all, as it's portrayed as a space-like dimension. This means that, very roughly, we should think time flows only if it's coherent to think of space flowing, which seems implausible.

On the plus side …

Perhaps we have reason to take heart. Although General Relativity plays a role in relevant experimental predictions, those predictions are actually coming from a particular theory of quantum gravity: string theory.

This theory reconciles our best theories of the very big (General Relativity) with the very small (quantum theory).

Some versions of string theory posit as many as 11 dimensions.

With so many, surely it'll be possible to make some sense of the idea that time flows … won't it?

The Higgs boson

Anna Phan, Postgraduate Student, University of Melbourne

Theoretical physics is full of mysteries and unknowns. In the case of some particles, we can predict their existence even if we can't find them.

Such is the case with the Higgs boson. And yet, detecting this particle would revolutionise physics as we know it.

Let's start with mass

When you place your suitcase on the scales at the airport check-in counter, you are hoping it weighs less than the limit so you won't have to pay any excess baggage fees.

The force of Earth's gravity acting on the suitcase's mass determines its weight. A suitcase that weighs 20 kg on Earth would weigh 3 kg on the moon, while its mass remains the same. What determines the suitcase's mass? And even more fundamentally, what is mass?

This is one of the most important questions in particle physics today. The leading explanation for the origin of mass is the Higgs mechanism developed in 1964, which involves the Higgs field and the Higgs boson.

My favourite description of the Higgs mechanism comes from David J. Miller, winner of a competition among physicists to find the best way of explaining the physics to the UK Science Minister in 1993 to acquire funding. The analogy goes something like this:

> Imagine that a room full of physicists chattering quietly is like space filled with the Higgs field. A well-known scientist walks in, creating a disturbance as he moves across the

room and attracting a group of admirers with every step. This increases his resistance to movement. In other words, he acquires mass, just like a particle moving through the Higgs field.

Now imagine if, instead of a well-known scientist entering, somebody started a rumour. As the rumour spreads throughout the room, it creates the same kind of grouping, but this time it's the scientists grouping together. It's these groups that are the Higgs bosons. If we find these groups, we can prove the Higgs field exists and thus explain the origin of mass.

What's at stake?

The discovery of the Higgs boson has vast and important consequences. It is the final missing piece of the Standard Model, the theory physicists use to describe the electromagnetic, strong and weak forces. All the other particles in the Standard Model have been proven to exist through experiment.

An example of such a prediction and subsequent discovery are the W and Z bosons, which mediate the weak force – the fundamental force that causes beta decay, a form of radioactivity. These particles were predicted in 1968 and discovered in 1983, an achievement so significant that Carlo Rubbia and Simon van der Meer were awarded a Nobel Prize in 1984.

Finding the Higgs boson will provide insight into why particles have certain masses, and will help to develop subsequent physics.

Why has it taken so long?

The technical problem is that the Standard Model doesn't predict the mass of the Higgs boson, which makes it more difficult to identify. Physicists have to look for it by systematically searching a range of mass within which it is predicted to exist.

Scientists at the previous LEP particle accelerator at CERN, the European Organization for Nuclear Research near Geneva, felt they came close before the machine shut down in 2000.

In July 2012, scientists using the Tevatron accelerator at Fermilab near Chicago saw hints of a new particle. These hints were confirmed by physicists using the Large Hadron Collider (LHC) accelerator at CERN, with the announcement of the discovery of a new Higgs-like particle making headlines worldwide.

By March 2013, the new particle had been shown to behave, interact and decay like a Higgs boson in many of the ways predicted by the Standard Model; leading CERN to claim it is a Higgs boson.

However, it remains an open question whether this is the Higgs boson of the Standard Model of particle physics, or one of several such bosons predicted in theories that go beyond the Standard Model.

Ultimately, upgraded and new accelerators will be needed to understand the interactions of the newly discovered Higgs boson at a deeper level, but for now, it looks as though we have an answer to the origin of mass.

What is an isotope?

Elizabeth Williams, Research Fellow in Nuclear Physics, ANU

If you've ever studied a periodic table of the elements you're probably already aware that it reveals a great deal about the chemical properties of the atoms that make up our world.

But you may not realise that each square on the periodic table actually represents a family of isotopes – atoms that generally share the same name and chemical properties, but have different masses.

To understand what isotopes are and how we can use them, we need to take a closer look at the interior of an atom.

Building blocks of matter

An atom is composed of an incredibly dense core (nucleus) of protons and neutrons, surrounded by a diffuse cloud of electrons.

You can think of protons and neutrons as the same kind of particle with one key difference: the protons are positively charged, while neutrons carry no charge. This means protons can 'feel' electric or magnetic fields, and neutrons cannot.

The electrons, which are much lighter than protons or neutrons, carry the same magnitude of charge as a proton but with the opposite sign, meaning each atom that has equal numbers of protons and electrons is electrically neutral.

It is the electrons that determine the chemical behaviour of a particular element.

Isotopes of an element share the same number of protons but have different numbers of neutrons. Let's use carbon as an example.

There are three isotopes of carbon found in nature: carbon-12, carbon-13, and carbon-14. All three have six protons, but their

PERIODIC TABLE OF THE ELEMENTS

H																	He
Li	Be											B	C	N	O	F	Ne
Na	Mg											Al	Si	P	S	Cl	Ar
K	Ca	Sc	Ti	V	Cr	Mn	Fe	Co	Ni	Cu	Zn	Ga	Ge	As	Se	Br	Kr
Rb	Sr	Y	Zr	Nb	Mo	Tc	Ru	Rh	Pd	Ag	Cd	In	Sn	Sb	Te	I	Xe
Cs	Ba	La-Lu	Hf	Ta	W	Re	Os	Ir	Pt	Au	Hg	Tl	Pb	Bi	Po	At	Rn
Fr	Ra	Ac-Lr	Rf	Db	Sg	Bh	Hs	Mt	Uun	Uuu	Uub	Uut	Uuq	Uup	Uuh	Uus	Uuo

Lanthanide series	La	Ce	Pr	Nd	Pm	Sm	Eu	Gd	Tb	Dy	Ho	Er	Tm	Yb	Lu
Actinide series	Ac	Th	Pa	U	Np	Pu	Am	Cm	Bk	Cf	Es	Fm	Md	No	Lr

The periodic table of elements. Jelena Zaric/Dreamstime

neutron numbers – 6, 7, and 8, respectively – differ. This means all three isotopes have different atomic masses (carbon-14 being the heaviest), but share the same atomic number ($Z = 6$).

Chemically, all three are indistinguishable, because the number of electrons in each of these three isotopes is the same.

So different isotopes of the same element are identical, chemically speaking. But some isotopes have the ability to circumvent this rule by transforming into another element entirely.

Marching towards stability

Some isotopes have this transformative ability because not all isotopes are stable. This is what led British chemist Frederick Soddy to his Nobel Prize-winning discovery of isotopes in 1913.

Some isotopes, such as carbon-12, will happily continue to exist as carbon unless something extraordinary happens. Others – carbon-14, say – will at some point decay into a stable isotope nearby.

In this case, one of the neutrons in carbon-14 changes into a proton, forming nitrogen-14. During this process, known as beta decay, the nucleus emits radiation in the form of an electron and an antineutrino (an antimatter particle).

There are many factors that can cause a nucleus to decay. One of the most important is the ratio of protons to neutrons a nucleus has. If a nucleus has too many neutrons (the definition of 'too many' depends on how heavy the nucleus is), there is a chance that it will decay towards stability.

The same is true if a nucleus has too many protons. This is one of the reasons why some isotopes of a given element are radioactive, while others are not.

From the bellies of stars

By now, you may be wondering how all these isotopes were created in the first place. As it turns out, this question is a complex one, but lends some truth to the adage that we are all made of stardust.

Some of the lighter isotopes were formed very early in the history of the universe, during the Big Bang. Others result from processes that happen within stars or as a result of chance collisions between highly energetic nuclei – known as cosmic rays – within our atmosphere.

Most naturally existing isotopes are the final (stable or long-lived) product resulting from a long series of nuclear reactions and decays.

In most of these cases, light nuclei have had to smash together with enough energy to allow the strong force – a glue-like bond that forms when protons and neutrons get close enough to touch – to overcome the electromagnetic force, which pushes protons apart. If the strong force wins out, the colliding nuclei bind together, or fuse, to form a heavier nucleus.

Our sun is a good example of this. One of its main sources of power is a series of fusion reactions and beta decay processes that transform hydrogen into helium.

Transforming knowledge into tools

Since the early 1900s, when the existence of isotopes was first realised, nuclear physicists and chemists have been seeking out ways to study how isotopes can be formed, how they decay, and how we might use them.

As it turns out, the nature of isotopes – their chemical uniformity, their nuclear distinctiveness – makes them useful for a wide range of applications in fields as diverse as medicine, archaeology, agriculture, power generation and mining.

If you have ever had a PET scan, you have benefited from a by-product of the radioactive decay of certain isotopes (often called medical isotopes). We produce these medical isotopes using our knowledge of how nuclear reactions proceed, with the help of nuclear reactors or accelerators called cyclotrons.

But we have also found ways to use naturally occurring radioactive isotopes. Carbon dating, for example, makes use of the long-lived isotope carbon-14 to determine how old objects are.

Under normal circumstances, carbon-14 is produced in our atmosphere via cosmic ray reactions with nitrogen-14. It has a half-life of roughly 5700 years, which means that half of a quantity of carbon-14 will have decayed away in that time.

While a biological organism is alive, it takes in approximately one carbon-14 isotope for every trillion stable carbon-12 isotopes and the carbon-12 to carbon-14 ratio stays about the same while the organism lives. Once it dies, new intake of carbon stops.

This means the ratio of carbon-14 to carbon-12 changes in the remains of this organism over time.

If we extract carbon using chemical methods from a sample, we can then apply a method called accelerator mass spectrometry (AMS) to separate out the individual carbon isotopes by weight.

AMS makes use of the fact that accelerated particles with the same charge but different masses follow separate paths through magnetic fields. By making use of these separate paths, we can determine isotope ratios with incredible accuracy.

As you can see from these examples, we apply our knowledge of isotopes in a variety of ways. We produce them, detect them, extract them, and study them with the dual purpose of understanding why the atomic nucleus behaves as it does, and how we can harness its power for our benefit.

What is Chaos Theory?

Jonathan Borwein, Laureate Professor of Mathematics, University of Newcastle
Michael Rose, PhD Candidate, School of Mathematical and Physical Sciences, University of Newcastle

'Clouds are not spheres, mountains are not cones ... Nature exhibits not simply a higher degree but an altogether different level of complexity.'

Benoît Mandelbröt, *The Fractal Geometry of Nature*

'Chaos (n): the inherent unpredictability in the behavior of a complex natural system.'

Merriam-Webster Dictionary

Chaos Theory is a delicious contradiction – a science of predicting the behaviour of 'inherently unpredictable' systems. It is a mathematical toolkit that allows us to extract beautifully ordered structures from a sea of chaos – a window into the complex workings of such diverse natural systems as the beating of the human heart and the trajectories of asteroids.

Welcome to one of the most marvellous fields of modern mathematics.

At the centre of Chaos Theory is the fascinating idea that order and chaos are not always diametrically opposed. Chaotic systems are an intimate mix of the two: from the outside they display unpredictable and chaotic behaviour, but expose the inner workings and you discover a perfectly deterministic set of equations ticking like clockwork. Some systems flip this premise around, with orderly effects emerging out of turbulent and chaotic causes.

How can order on a small scale produce chaos on a larger scale? And how can we tell the difference between pure randomness and orderly patterns that are cloaked in chaos? The answers can be found in three common features shared by most chaotic systems.

Butterflies make all the difference

'Tiny variations ... never repeat, and vastly affect the outcome.'

Jeff Goldblum ('Ian Malcolm'), *Jurassic Park*

In 1961, meteorologist Edward Lorenz made a profound discovery.

Lorenz was using the new-found power of computers in an attempt to more accurately predict the weather. He created a mathematical model which, when supplied with a set of numbers representing the current weather, could predict the weather a few minutes in advance. Once this program was up and running, Lorenz could produce long-term forecasts by feeding the predicted weather back into the computer over and over again, with each run forecasting further into the future. Accurate minute-by-minute forecasts added up into days, and then weeks.

One day, Lorenz decided to rerun one of his forecasts. To save time, he decided not to start from scratch; instead he took the computer's prediction from halfway through the first run and used that as the starting point. After a coffee break, he returned to discover something unexpected. Although the computer's new predictions started out the same as before, the two sets of predictions soon began diverging drastically. What had gone wrong?

Lorenz realised the computer was printing out the predictions to three decimal places, but actually crunching the numbers internally using six decimal places. So while Lorenz had started the second run with the number 0.506, the original run used 0.506127. A difference of one part in a thousand: the same sort of difference that a flap of a butterfly's wing might make to the breeze on your face. The starting weather conditions had been virtually identical. The two predictions were anything but.

Lorenz had found the seeds of chaos. In systems that behave nicely – without chaotic effects – small differences only produce small effects. In this case, Lorenz's equations were causing errors to steadily grow over time. This meant that tiny errors in the measurement of the current weather would not stay tiny, but relentlessly increased in size each time they were fed back into the computer until they had completely swamped the predictions. Lorenz famously illustrated this effect with the analogy of a butterfly flapping its wings and thereby causing the formation of a hurricane half a world away.

A nice way to see this 'butterfly effect' for yourself is with a game of pool or billiards. No matter how consistent you are with the first shot (the break), the smallest difference in the speed and angle with which you strike the white ball will cause the pack of billiards to scatter in wildly different directions every time. The smallest of differences are producing large effects – the hallmark of a chaotic system.

It is worth noting that the laws of physics that determine how the billiard balls move are precise and unambiguous: they allow no room for randomness. What at first glance appears to be random behaviour is completely deterministic – it only seems random because imperceptible changes are making all the difference. The rate at which these tiny differences stack up provides each chaotic system with a prediction horizon – a length of time beyond which we can no longer accurately forecast its behaviour. In the case of the weather, the prediction horizon is nowadays about one week (thanks to ever-improving measuring instruments and models). Fifty years ago it was 18 hours. Two weeks is believed to be the limit we could ever achieve, however much better computers and software get.

Surprisingly, the solar system is a chaotic system too – with a prediction horizon of a hundred million years. It was the first chaotic system to be discovered, long before there was a Chaos Theory.

In 1887, French mathematician Henri Poincaré showed that while Newton's theory of gravity could perfectly predict how two planetary bodies would orbit under their mutual attraction, adding a third body to the mix rendered the equations unsolvable. The best we can do for three bodies is to predict their movements moment by moment, and feed those predictions back into our equations ...

Though the dance of the planets has a lengthy prediction horizon, the effects of chaos cannot be ignored, for the intricate interplay of gravitational tugs among the planets has a large influence on the trajectories of the asteroids. Keeping an eye on the asteroids is difficult but worthwhile, since such chaotic effects may one day fling an unwelcome surprise our way. On the flip side, they can also divert external surprises, such as steering comets away from a potential collision with Earth.

Attractive, strange behaviour

> 'First, a highly regular motion towards the attractor ... then a much more irregular motion on it.'
>
> Ian Stewart, *The Magical Maze*

Stability is desirable in many scenarios, such as flying. Commercial aircraft are aerodynamically stable, so a small turbulent nudge (possibly butterfly-related) won't push the plane out of a level flight path. Comfortingly, it takes a large change in the flight controls to effect a large change in the plane's motion.

On the other hand, this stability is an inconvenience to fighter pilots, who prefer their aircraft to make rapid changes with minimal effort. Modern fighter jets achieve great manoeuvrability by being aerodynamically unstable – the slightest nudge is enough to drastically alter their flight path. Consequently, they are equipped with on-board computers which constantly and delicately adjust the flight surfaces to cancel out the unwanted butterfly effects, leaving the pilot free to exploit his own.

If you can tease out the pattern's underlying chaotic systems, you can gain a measure of control over randomness and turn instability into an asset. The key to unlocking the hidden structure of a chaotic system is in determining its preferred set of behaviours – known to mathematicians as its attractor. Mathematician Ian Stewart uses this example to illustrate an attractor:

> *Imagine taking a ping-pong ball far out into the ocean and letting it go. If released above the water it will fall, and if released underwater it will float. No matter where it starts, the ball will immediately move in a very predictable way towards its attractor – the ocean surface. Once there it clings to its attractor as it is buffeted to and fro in a literal sea of chaos, and quickly moves back to the surface if temporarily thrown above or dumped below the waves.*

Though we may not be able to predict exactly how a chaotic system will behave moment to moment, knowing the attractor allows us to narrow down the possibilities. It also allows us to accurately predict how the system will respond if it is jolted off its attractor.

Mathematicians use the concept of a 'phase space' – a space in which all possible states of a system are represented – to describe the possible behaviours of a system geometrically. Phase space is not (always) like regular space – each location in phase space corresponds to a different configuration of the system. The behaviour of the system can be observed by placing a point at the location representing the starting configuration and watching how that point moves through the phase space.

In phase space, a stable system will move predictably towards a very simple attractor (which will look like a single point in the phase space if the system settles down, or a simple loop if the system cycles between different configurations repeatedly). A chaotic system will also move predictably towards its attractor in phase space – but instead of points or simple loops, we see 'strange attractors' appear – complex and beautiful shapes (known as fractals) that twist and turn, intricately detailed at all possible scales. The branch of fractal mathematics, pioneered by the French-American mathematician Benoît Mandelbröt, allows us to come to grips with the preferred behaviour of this system, even as the incredibly intricate shape of the attractor prevents us from predicting exactly how the system will evolve once it reaches it.

Phase space may seem fairly abstract, but one important application lies in understanding your heartbeat. The millions of cells that make up your heart are constantly contracting and relaxing separately as part of an intricate chaotic system with complicated attractors. These millions of cells must work in sync, contracting in just the right sequence at just the right time to produce a healthy heartbeat. Fortunately, this intricate state of synchronisation is an attractor of the system – but it is not the only one. If the system is jolted somehow, it may find itself on an altogether different attractor called fibrillation, in which the cells constantly contract and relax in the wrong sequence. The purpose of a defibrillator – the device that applies a large voltage of electricity across the heart – is not to 'restart' the heart cells, but to give the chaotic system enough of a kick to move it off the fibrillating attractor and back to the healthy heartbeat attractor.

The main benefit to having a chaotic heart is that tiny variations in the way those millions of cells contract serve to distribute the load more evenly, reducing wear and tear on your heart and allowing it to pump decades longer than would otherwise be possible.

The cascade into chaos

'Universality made the difference between beautiful and useful.'

James Gleick, *Chaos*

Chaos Theory is not solely the province of mathematicians. It is notable for drawing together specialists from many diverse fields – physicists and biologists, computer scientists and economists. Not only can chaotic systems be found almost anywhere you care to look, they share many common features independently of where they came from.

Consider both a dripping tap and the supercooled liquid helium the Large Hadron Collider uses as a coolant (making parts of the LHC colder than deep space). Both are non-chaotic systems – at first – but as you slowly heat the helium, tiny convection cells begin to form, and as you slowly open the tap, the dripping sounds will change in character. Eventually the increases in temperature and water flow will cascade into the chaos of boiling helium and rushing water, respectively.

Amazingly, the transition from order to chaos in these systems is controlled by the exact same number – the Feigenbaum constant, which indicates that all chaotic systems of this type will move from order to chaos at the same rate.

From dripping taps to the LHC, from a beating heart to the dance of the planets, chaos is all around us. Chaos Theory has turned everyone's attention back to things we once thought we understood, and shown us that nature is far more complex and surprising than we had ever imagined.

What is wave-particle duality?

Tim Davis, *Principal Research Scientist, Materials Science and Engineering, CSIRO*

Our notion of reality is built on everyday experiences. But wave-particle duality is so strange that we are forced to re-examine our common conceptions.

Wave-particle duality refers to the fundamental property of matter where, at one moment it appears like a wave, and yet at another moment it acts like a particle.

To understand wave-particle duality it's worth looking at differences between particles and waves.

We are all familiar with particles, whether they are marbles, grains of sand, salt in a salt-shaker, atoms, electrons, and so on.

The properties of particles can be demonstrated with a marble. The marble is a spherical lump of glass located at some point in space. If we flick the marble with our finger, we impart energy to it. This is kinetic – moving – energy, and the moving marble takes this energy with it. Marbles thrown in the air come crashing down, each marble imparting energy where it strikes the floor.

In contrast, waves are spread out. Examples of waves are the big rollers on the open ocean, ripples in a pond, sound waves and light waves.

If at one moment the wave is localised, some time later it will have spread out over a large region, like the ripples when we drop a pebble in a pond. The wave carries with it energy related to its motion. Unlike the particle, the energy is distributed over space because the wave is spread out.

Why waves are so different from particles

Colliding particles will bounce off each other but colliding waves pass through one another and emerge unchanged. But overlap-

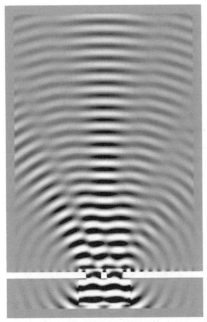

The interference pattern of a wave incident on two holes in a screen. The holes can be seen near the bottom of the image as two breaks in the white line. The waves above the screen show regions of destructive interference, where the wave crests overlap troughs and cancel out, and regions of constructive interference, where the wave crests overlap crests and reinforce. Tim Davis

ping waves can interfere – where a trough overlaps a crest the wave can disappear altogether.

This can be seen when parts of a wave pass through closely spaced holes in a screen. The waves spread out in all directions and interfere, leading to regions in space where the wave disappears and regions where it becomes stronger.

The image above shows an example of the double slit experiment invented by English polymath Thomas Young. This phenomenon is called diffraction.

In contrast, a marble thrown at the screen either bounces off or goes straight through one of the holes. On the other side of the screen, the marble will be found travelling in one of two directions, depending on which hole it went through.

Wave goodbye to waves

The phenomenon of diffraction is a well-known property of light waves. But at the beginning of the 20th century, a problem was found with the theories of light waves emitted from hot objects, such as hot coals in a fire or light from the sun.

This light is called black-body radiation. These theories would always predict infinite energy for the light emitted beyond the blue end of the spectrum. This was known as the ultraviolet catastrophe.

The answer was to assume the energy of light waves was not continuous but came in fixed amounts, as if it was composed of a large number of particles, like our handful of marbles. So the notion came about that light waves act like particles. These particles are called photons.

If light, that we thought was wave-like, also behaves like a particle, could it be that objects such as electrons and atoms, that are particle-like, can behave like waves?

To explain the structure and behaviour of atoms it was thought necessary to assume that particles have wave-like properties. If this is true, a particle should diffract through a pair of closely spaced holes, just like a wave.

Electron and atom diffraction

Experiments proved atomic particles act just like waves. When we fire electrons at one side of a screen with two closely spaced holes and measure the distribution of electrons on the other side, we don't see two peaks, one for each hole, but a complete diffraction pattern, just as if we had been using waves.

This is another example of the Young's slit experiment we showed above, but this time using electron waves. These notions form the basis of quantum theory – perhaps the most successful theory scientists have ever developed.

The bizarre thing about the diffraction experiment is that the electron wave doesn't deposit energy over the entire surface of the detector, as you might expect with a wave crashing on the shore.

The energy of the electron is deposited at a point, just as if it were a particle. So while the electron propagates through space like a wave, it interacts at a point like a particle. This is known as wave-particle duality.

It moves in mysterious waves

If the electron or photon propagates as a wave but deposits its energy at a point, what happens to the rest of the wave?

It disappears, from all over space, never to be seen again! Somehow, those parts of the wave distant from the point of inter-action know that the energy has been lost and disappear, instantaneously.

If this happened with ocean waves, one of the surfers on the wave would receive all the energy and at that moment the ocean wave would disappear, all along the length of the beach. One surfer would be shooting along the surface of the water and the rest would be sitting becalmed on the surface.

This is what happens with photons, electrons and even atom waves. Naturally enough, this conundrum upset a lot of scientists, Einstein included. It is usually swept under the carpet and glibly referred to as 'the collapse of the wave function' on measurement.

Certain uncertainty

As the wave propagates, where is the particle? We don't know for sure. It is located somewhere in the region of space with a dimen-sion similar to the distribution of wavelengths that define its wave. This is known as Heisenberg's uncertainty principle.

For common everyday particles, such as marbles, salt and sand, their wavelengths are so small that their location can be accurately measured. For atoms and electrons, this becomes less clear.

In the diffraction experiment the electron wavelength is large, so the location of the electron is very uncertain. The electron actually travels through both slits at once, just like a wave. In terms of particles it becomes impossible for us to really imagine this because it conflicts with everyday experience.

Einstein worried about where the particle is located and decided information was missing in the quantum theory. In a celebrated paper on hidden variables, Einstein and his colleagues Nathan Rosen and Boris Podolsky derived two alternatives: either quantum theory was wrong or the problem resided in our notion of reality itself.

A series of precise and clever experiments proved quantum theory was correct. Our notion of reality *is* at fault.

Ghostly behaviour

But this is not the end of the story. The experiments that disproved our notions of reality involved two particles linked together as a single wave. Measurements on one particle affect the physical properties of the other particle, even though they can be far apart. This is known as 'spooky action at a distance' and is a consequence of quantum entanglement. This occurs when two particles act on one another and become an entangled system. When a pair or group of particles can only be described as a group with one quantum state, rather than discrete quantum states put together, we say the particles are 'entangled'.

It is a very subtle concept, but it forms the basis of quantum computers and quantum cryptography.

So what's wrong with reality?

At this point the whole problem gets very difficult to get your mind around. But don't get too worried about this. As Richard Feynman said: 'I think I can safely say that nobody understands quantum mechanics.'

Most people working in this field just get used to the concept and get on with their lives, or become philosophers.

And as for reality?

I think Professor Feynman has the last word on that one, too: '... the *paradox* is only a conflict between reality and your feeling of what reality ought to be.'

GLOSSARY

Atomic clock: an extremely precise clock whose rate is controlled by a periodic process (such as vibration, or the absorption or emission of electromagnetic radiation) that occurs at a steady rate in atoms or molecules. The standard atomic clock is based on the vibrations of cesium atoms.

Bacterium: any member of a large group of one-celled organisms that lack a cell nucleus, reproduce by fission or by forming spores, and in some cases cause disease. They are the most abundant life forms on Earth, found in all living things and in all environments. Bacteria usually live off other organisms.

Base pair: any of the pairs of nucleotides connecting the complementary strands of a molecule of DNA. The base pairs are adenine-thymine and guanine-cytosine. Base pairs are the 'rungs' of the DNA ladder.

Boson: any of a class of elementary or composite particles that can potentially be in the same quantum state.

Catalyst: substance that starts or speeds up a chemical reaction while not being permanently changed itself.

Chromosome: a structure in all living cells, consisting of a single molecule of DNA bonded to various proteins and carrying the genes that determine heredity. In multicellular organisms, chromosomes occur in pairs in all the cells except the reproductive cells, which have one of each chromosome, and in some red blood cells.

Colour charge: the property of quarks and gluons that determines their strong force interaction with each other (especially the attractive and repulsive forces between them). Colour charge is considered to be a form of charge, much like electrical charge. There are three basic colour charges (blue, green, and red) and associated anti-colours. Colour charge has nothing to do with colours seen by the eye.

Cyclotron: a type of particle accelerator that accelerates charged subatomic particles, such as protons and electrons, in an outward spiral, greatly increasing their energies. Cyclotrons are used to bring about high-speed particle collisions in order to study subatomic structures.

Cytoplasm: a jelly-like material that constitutes much of a cell inside the cell membrane, and surrounds the nucleus.

Diffraction: bending and spreading of a wave, e.g. a light wave, around the edge of an object.

DNA: deoxyribonucleic acid. The genetic material that determines the makeup of all living cells and many viruses. It consists of two long strands of nucleotides linked together in a structure resembling a ladder twisted into a spiral. In multicellular organisms, the DNA is contained in the nucleus (bound to proteins known as histones) and in mitochondria and chloroplasts.

Ecosystem: a community of organisms and their physical environment, viewed as a system of interacting and interdependent relationships. It includes interactions between both living and non-living components of the system.

Epigenome: the total set of chemical compounds that modify or mark the genome in a way that determines how the gene is expressed. Different cells have different epigenetic marks. They are not part of the DNA itself, but can be passed on from cell to cell as cells divide, and from one generation to the next.

Expression (of a gene): the process of interpreting the genetic code stored in DNA. The properties of the expression determine the organism's phenotype.

Fermion: An elementary or composite particle, such as an electron, quark or proton, whose spin is an integer multiple of ½. Fermions act on each other by exchanging bosons. No two fermions can exist in the same quantum state.

Flavour (quarks): a way of referring to the type of quark. Either up quark, down quark, top quark, bottom quark, charm quark and strange quark. It has no relation to the sense of taste.

Fractal: a geometric pattern that repeats at ever-smaller scales to produce irregular shapes and surfaces that cannot be represented by classical geometry. Fractals are used especially in computer modelling of irregular patterns and structures in nature.

Genome: the entire amount of genetic information in the chromosomes of an organism, including genes and DNA sequences.

Gluon: subatomic particle that mediates the strong force. An exchange of gluons between two quarks changes the colour of the quarks and results in the attractive force holding them together in hadrons. Gluons are bosons.

Hadron: any of a class of subatomic particles composed of a combination of two or more quarks or antiquarks. Quarks (and antiquarks) of different colours are held together in hadrons by the strong nuclear force. Hadrons include both baryons (composed of three quarks or three antiquarks) and mesons (composed of a quark and an antiquark). The combination of quark colours in a hadron must be neutral, e.g. red and anti-red, or red, blue and green. Note that the term 'colour' does not describe an actual shade.

Hypothesis: a proposition that attempts to explain a set of facts in a unified way. It generally forms the basis of experiments designed to establish its plausibility. A hypothesis can never be proven to be unequivocally true, but can sometimes be verified beyond reasonable doubt in the context of a particular theoretical approach.

Law: a statement that describes invariable relationships among phenomena under a specified set of conditions.

Lepton: any member of a family of elementary particles that interact through the weak force and do not participate in the strong force. Leptons include electrons, muons, tau particles, and their respective neutrinos, the electron neutrino, the muon neutrino and the tau neutrino. The antiparticles of these six particles are also leptons.

Meson: any of a family of subatomic particles composed of a quark and an antiquark. Their masses are generally somewhere between leptons and baryons, and they can have positive, negative or neutral charge. They are a subclass of hadrons.

Microsecond: one millionth of a second.

Mitochondria: structures in the cytoplasm of all cells except bacteria, in which food molecules (sugars, fatty acids, and amino acids) are broken down in the presence of oxygen and converted to energy. They have an inner and outer membrane. The inner membrane encloses a liquid that contains DNA, RNA, small ribosomes, and solutes. The DNA in mitochondria is genetically distinct from that in the cell nucleus, and mitochondria can manufacture some of their own proteins independently of the rest of the cell. Mitochondria were probably originally separate, single-celled organisms that became so symbiotic with their hosts as to be indispensible. Mitochondrial DNA is considered to be a remnant of a past existence as a separate organism.

Muon: an elementary particle with a negative charge and a half-life of 2 microseconds. Muons decay into electrons and neutrinos/antineutrinos.

Mutation: a change in the structure of the genes or chromosomes of an organism. Mutations occurring in the reproductive cells, such as an egg or sperm, can be passed from one generation to the next. Most mutations occur in junk DNA and have no discernible effects on an organism's survival. Other mutations mostly have harmful effects, while some can increase an organism's ability to survive.

Nanometre: one billionth of a metre.

Neutrino: Any of three electrically neutral subatomic particles with extremely low mass. These include the electron-neutrino, the muon-neutrino, and the tau-neutrino. The study of neutrinos in cosmic rays suggests that neutrinos can transform into each other. For this phenomenon to be theoretically possible, the three neutrinos must have distinct masses.

Nucleotide: any of a group of molecules that link together, forming the building blocks of DNA or RNA: composed of a phosphate group, the bases adenine, cytosine, guanine and thymine, and a sugar. In RNA the thymine base is replaced by uracil.

Nucleus (biology): the part of the cell (in multicellular organisms) that contains nearly all the cell's DNA and controls its metabolism, growth, and reproduction.

Nucleus (physics): the positively charged central region of an atom, composed of one or more protons and (for all atoms except hydrogen) one or more neutrons, containing most of the mass of the atom.

Particle accelerator: machines that increase the speed and energy of protons, electrons or other atomic particles, and direct them at atomic nuclei or other particles to cause high-energy collisions. These collisions produce yet other particles, whose paths are then tracked and analysed. Particle accelerators are used to study the nature of the atomic nucleus, subatomic particles, and the forces relating them, and to create radioactive isotopes. Less complex devices such as cathode ray tubes are also particle accelerators.

Phase space: a hypothetical space constructed so as to have as many coordinates as are needed to define the state of a given substance or system.

Phenotype: the physical appearance of an organism as opposed to its genetic makeup. The phenotype of an organism depends both on which genes are dominant and on the interaction between genes and environment.

Photon: subatomic particle that carries the electromagnetic force. It has a rest mass of zero, but has measurable momentum, can be deflected by a gravitational field, and can exert a force. It has no electric charge, has an indefinitely long lifetime, and is its own antiparticle.

Photosynthesis: process by which green plants, algae, diatoms and some types of bacteria make carbohydrates from carbon dioxide and water in the presence of chlorophyll, using energy captured from sunlight by chlorophyll, and releasing excess oxygen as a by-product.

Photovoltaic: capable of producing a voltage when exposed to radiant energy, especially light.

Probabilistic: subject to or involving chance variation – an essential concept in quantum mechanics.

Proteome: the complete set of proteins that the genetic material of an organism can express.

Pyrolysis: the chemical process of decomposition under the effect of heat.

Quantum mechanics: a fundamental theory of matter and energy that accommodates facts previous physical theories could not account for. Among these are that energy is absorbed and released in small, discrete quantities (quanta), and that all matter displays both wave-like and particle-like properties, especially at atomic and subatomic scales. Quantum mechanics suggests that the behaviour of matter and energy is not deterministic and the effect of the observer on the physical system being observed must be understood as a part of that system.

Quark: any member of a group of elementary particles believed to be the fundamental units that combine to make up the subatomic particles known as hadrons (baryons, such as neutrons and protons, and mesons). There are six different flavours (or types) of quark: up quark, down quark, top quark, bottom quark, charm quark, and strange quark. Quarks have fractional electric charges, such as ⅓ the charge of an electron.

Randomised controlled (or control) trial (RCT): a study in which people are allocated at random to receive one of several clinical interventions. One of these interventions is the standard of comparison or control – standard practice, a placebo or no intervention at all. RCTs seek to measure and compare the outcomes that follow the interventions. Because the outcomes are measured, RCTs are quantitative studies.

Ribosome: a sphere-shaped structure within the cytoplasm of a cell. It is composed of RNA and protein and is the site of protein synthesis.

RNA: ribonucleic acid. It is the nucleic acid used in key metabolic processes in all living cells and carries the genetic information of many viruses. Unlike DNA, RNA consists of a single strand, and occurs in a variety of lengths and shapes. In multicellular organisms, RNA is produced in the cell nucleus. There are several types of RNA:
- mRNA carries genetic information from the cell nucleus to the structures in the cytoplasm (known as ribosomes) where protein synthesis takes place.
- rRNA is the main structural component of the ribosome.
- tRNA delivers the amino acids necessary for protein synthesis to the ribosomes.

Scintigraphy: the production of two-dimensional images of the distribution of radioactivity in tissues after the internal administration of a radiopharmaceutical imaging agent. The images are obtained using a scintillation camera, which detects gamma radiation.

Sequence (DNA, RNA): the order in which subunits appear in a chain, such as amino acids in a polypeptide or nucleotide bases in a DNA or RNA molecule.

Sequencing (DNA, RNA): laboratory technique used to find out the sequence of nucleotide bases in a DNA or RNA molecule or fragment.

Strange attractor: a stable state or behaviour exhibited by some dynamic systems, especially turbulent ones, that can be represented as a nonrepeating pattern in the system's phase space.

Strong force: the fundamental force that mediates interactions between particles with colour charge, such as quarks and gluons. The strong force binds quarks together to form baryons such as protons and neutrons, maintains the binding of protons and neutrons together in atomic nuclei, and is responsible for many particle decay processes.

Tau particle (tauon): an elementary particle of the lepton family, with a mass about 3550 times that of the electron, a negative electric charge, and a lifetime of about 17 microseconds.

Theory: a set of statements or principles that explains a group of facts or phenomena. Most theories accepted by scientists have been repeatedly tested experimentally and can be used to make predictions about natural phenomena.

Transcribe (genetics): to copy genetic material from a strand of DNA to a complementary strand of RNA (messenger RNA). In multicellular organisms, transcription takes place in the nucleus before messenger RNA is transported to the ribosomes for protein synthesis.

Translate (genetics): to direct the assembly of a sequence of amino acids to make a protein. The process uses messenger RNA and takes place in the ribosomes.

Vaccine: a preparation of a weakened or killed pathogen, such as a bacterium or virus, or of part of the pathogen's structure, that stimulates immune cells to recognise and attack it, especially through antibody production.

Virus: extremely small, often disease-causing agents consisting of a particle containing a segment of RNA or DNA within a protein coating. Viruses are not technically considered living organisms: they neither breathe, metabolise nor reproduce on their own, but need a living host cell to make more viruses. Viruses reproduce first either by injecting their genetic material into the host cell or by fully entering the cell and shedding their protein coat. Rather than being primordial forms of life, viruses probably evolved from rogue pieces of cellular nucleic acids.

Weak force: the fundamental force that acts between leptons and is involved in the decay of hadrons. The weak nuclear force is responsible for nuclear beta decay (by changing the flavour of quarks) and for neutrino absorption and emission. It is weaker than either the strong nuclear force or the electromagnetic force, but stronger than gravity.